Synthesis Lectures on Computer Science

The series publishes short books on general computer science topics that will appeal to advanced students, researchers, and practitioners in a variety of areas within computer science.

Leonidas Anthopoulos
Editor

Smart City Standardization

A Global Perspective

Editor
Leonidas Anthopoulos
Department of Business Administration
University of Thessaly
Trikala, Greece

ISSN 1932-1228　　　　　　　ISSN 1932-1686　(electronic)
Synthesis Lectures on Computer Science
ISBN 978-3-031-95958-5　　　ISBN 978-3-031-95959-2　(eBook)
https://doi.org/10.1007/978-3-031-95959-2

© The Editor(s) (if applicable) and The Author(s), under exclusive license to Springer Nature Switzerland AG 2025

This work is subject to copyright. All rights are solely and exclusively licensed by the Publisher, whether the whole or part of the material is concerned, specifically the rights of translation, reprinting, reuse of illustrations, recitation, broadcasting, reproduction on microfilms or in any other physical way, and transmission or information storage and retrieval, electronic adaptation, computer software, or by similar or dissimilar methodology now known or hereafter developed.
The use of general descriptive names, registered names, trademarks, service marks, etc. in this publication does not imply, even in the absence of a specific statement, that such names are exempt from the relevant protective laws and regulations and therefore free for general use.
The publisher, the authors and the editors are safe to assume that the advice and information in this book are believed to be true and accurate at the date of publication. Neither the publisher nor the authors or the editors give a warranty, expressed or implied, with respect to the material contained herein or for any errors or omissions that may have been made. The publisher remains neutral with regard to jurisdictional claims in published maps and institutional affiliations.

This Springer imprint is published by the registered company Springer Nature Switzerland AG
The registered company address is: Gewerbestrasse 11, 6330 Cham, Switzerland

If disposing of this product, please recycle the paper.

Foreword

The 21st century is witnessing an unprecedented urbanization surge. Cities, as the epicenters of economic growth and innovation, are grappling with complex challenges such as climate change, inequality, and resource scarcity. To navigate these complexities and build sustainable, resilient, and inclusive urban environments, a standardized approach is imperative.

This book, *Smart City Standardization: A Global Perspective*, is a timely exploration of the critical role that standardization plays in shaping the future of our cities. It explores the multifaceted landscape of smart city standards, from global frameworks to regional and national initiatives.

At the heart of this endeavour is the United for Smart Sustainable Cities (U4SSC) initiative, a collaborative effort by 19 UN bodies, that provides cities worldwide with the tools and guidance needed to embark on their smart city journeys. By focusing on Key Performance Indicators (KPIs), U4SSC empowers cities to measure progress, identify areas for improvement, and ultimately enhance the quality of life for their citizens through building smarter, greener urban environments.

The International Telecommunication Union (ITU) has also emerged as a pivotal player in the standardization arena. By fostering collaboration among 193 UN nations, the ITU is driving the digital transformation of cities and laying the groundwork for a more connected and equitable world.

Beyond the international stage, this book delves into specific case studies of smart city standardization initiatives across the globe. From the United States' private sector-driven model to Spain's emphasis on smart tourism, and from China's focus on interoperable platforms to Japan's commitment to disaster resilience, this book offers a comprehensive exploration of diverse approaches to smart city standardization. The readers are also introduced to innovative projects such as France's Proxi-Produit, a proximity-based citizen service utilizing IoT, and Brazil's intelligent platform, a sophisticated system for assessing and improving smart city development across thousands of municipalities.

I have had the privilege of collaborating closely with Prof. Leonidas Anthopoulos at the International Telecommunication Union (ITU), the UN's specialized agency for ICTs, on the pre-standardization of CitiVerse, a new UN framework for citizen-centric future smart cities leveraging cutting-edge technologies which leads to international standards. Professor Anthopoulos' profound knowledge of the global smart city standardization landscape has consistently impressed me.

His new book on *Smart City Standardization: A Global Perspective* is a vital resource for policymakers, urban planners, technologists, and anyone interested in the future of our cities. By understanding the global landscape of standardization, we can accelerate the development of smart, sustainable, and inclusive urban environments that benefit all.

I am confident that this work will contribute to the development of a global framework for smart city standardization, ultimately leading to more sustainable, resilient, and equitable urban futures. Let us embark on this journey together to build a better tomorrow for all citizens.

Dr. Christina Yan Zhang
CEO of The Metaverse Institute
Co-chairman of the ITU Task Group on
Pre-standardization for the CitiVerse
London, UK
christina@metaverse-institute.org

Contents

1 **Introduction: Navigating the Landscape of Standardisation for Smart Cities** 1
Constantinos Marios Angelopoulos and Ramy Fathy

Part I The Context of Smart City Standards

2 **What KPIs Can Reveal and Help Achieve** 11
Barbara Kolm and Victoria Schmid

Part II Smart City Standardization: International Standards

3 **ITU Standards for Smart Cities: Driving Digital Transformation** 35
Hyoung Jun Kim

Part III Smart City Standardization: European Efforts

4 **ETSI Standards for Smart Cities: Standards for Interoperable, Sustainable and Accessible Citizen-Centric Services** 53
Laure Pourcin

Part IV Smart City Standardization: National Efforts

5 **United States Approach to Standards for Smart Cities** 75
W. Michael Dunaway and Cheyney M. O'Fallon

6 **Smart City Standards in Canada (SCSC)** 101
Val Wise

7	**The Spanish Vision of Smart Cities and Smart Tourist Destinations Through Standards**	113
	Ramón Ferri Tormo, Beatriz de Esteban Martín, and Tania Marcos Paramio	
8	**Smart City Standardization in Japan**	133
	David N. Nguyen, Yasuhiro Okuda, and Takeshi Furuno	
9	**Smart City Standardization in Greece**	153
	Panagiotis Karadimos and Leonidas Anthopoulos	

Part V Cases of Defining or Applying Smart City Standards

10	**Brazilian Smart City Standards—Challenges and Opportunities in the Adaptation and Expansion of the SSCMM-ITU: Platform inteli.gente Management and Governance System for Digital Transformation and Sustainable Development**	181
	Luísa Paseto, Márcia Regina Martins Martinez, Ricarda Carolina Rende, Rodrigo Barbosa Paula, and Andre Carlos Ponce de Leon Ferreira de Carvalho	
11	**Chinese Smart City Standards**	197
	Ziqin Sang and Wenying Du	
12	**Smart City Standardization in France: The Case of Proxi-Produit Project to the DPP Policy**	217
	Sandoche Balakrichenan	
13	**Conclusions: Building a Cohesive Future Through Smart City Standards**	235
	Okan Geray	

Introduction: Navigating the Landscape of Standardisation for Smart Cities

Constantinos Marios Angelopoulos and Ramy Fathy

Abstract

Cities are the cradle of human societies and modern civilization. Following the agricultural revolution in circa 10,000 BC, the emergence of the first cities marked the transition from hunting and gathering communities to settled societies.

1.1 Smart Cities and Standardization

Cities are the cradle of human societies and modern civilization. Following the agricultural revolution in circa 10,000 BC, the emergence of the first cities marked the transition from hunting and gathering communities to settled societies. Depending on the socio-economic and geo-political environments, cities around the globe have assumed different development paths. However, in all cases, cities have been driving progress in terms of governance, commerce and economic growth, culture, science and technology, and every other facet of human societies.

Along with the increase in human population and the ever-increasing interconnectedness of societies, cities also increase in size, further amplifying their importance. In their report "Metropolitan Areas" (OECD 2015), OECD review the impact of 281 metropolitan

C. M. Angelopoulos (✉)
Bournemouth University, Poole, UK
e-mail: mangelopoulos@Bournemouth.ac.uk

International Hellenic University, Thessaloniki, Greece

R. Fathy
National Telecom Regulatory Authority (NTRA), Nasr City, Egypt
e-mail: ramy.ahmed@ieee.org

© The Author(s), under exclusive license to Springer Nature Switzerland AG 2025
L. Anthopoulos (ed.), *Smart City Standardization*, Synthesis Lectures on Computer Science, https://doi.org/10.1007/978-3-031-95959-2_1

areas and find that 49% of the OECD population lived in that areas, generated 57% of gross domestic product (GDP) and 51% of employment in the OECD area. The report also found that metropolitan areas tend to be more productive than the rest of the economy, with the productivity gap (GDP per worker) being on the average 30% (OECD 2015). More recently, Eurostat came to similar findings regarding the impact of capital cities and metropolitan regions in European Union; their combined gross domestic product (GDP) was €3.1 trillion in 2019, which equated to 22.6% of the EU total (Eurostat 2024).

The capacity of cities to deliver this significant impact over time is largely driven by novel processes of governance and organisation, often underpinned by scientific and technological breakthroughs. More efficient urban planning, development of power and transport networks, and effective waste management are few indicative examples. In recent years, disruptive emerging ICTs have paved the way for the digitalisation of most of these processes. Digital systems and networks now enable the collection, curation, and processing of dense data in both space and time domains. This in turn enables better-informed process design and policy making at shorter and more frequent development cycles. The impact of emerging ICTs has been such, that a new paradigm has emerged; the paradigm of a Smart City.

Given the unique profile of each city and the highly diverse and fast evolving landscape of ICTs, how to effectively on-board new digital technologies is not a straightforward exercise. Each city has its own priorities and faces different challenges, and it may have access to a different set of digital technologies depending on the ambient socio-economic and geo-political environments. Furthermore, there are multiple stakeholders involved, often with competing interests—public administration, industry, academia, NGOs, etc.— among which compromises need to be reached more often than not. International standards developing organizations (SDO) constitute important *fora* for city stakeholders to exchange lessons learnt, disseminate good practices, and develop a common vision and understanding that will further facilitate the delivery of positive impact to citizens via economic growth and improvement of quality of life.

In the following, we focus on the domain of Smart Cities and first discuss how the corresponding standardisation landscape has been shaped by emerging ICT to now also include additional stakeholders. We also share our experiences from developing an international standard in the domain of Smart Cities underpinned by research in ICT. Apart from the technical work, the process also involved navigating the regulations governing the SDO, as well as the relationship dynamics among a diverse corpus of stakeholders for obtaining consent for the adoption of the standard. This chapter aims to serve as a gentle pre-ample to the remaining chapters of this tome presenting standardisation activities from several regions of the world in the Smart Cities domain.

1.2 Standardization "Within", "Outward", and "Beyond"

The development of widely adopted technical standards is a key mechanism for technology commercialization and economic development. Technical standards are the main vehicle supporting interoperability among different technologies and their implementations, thus greatly facilitating the establishment of a common reference framework for the involved stakeholders. This common reference creates the stable environment needed for technologies under research to be successfully commercialized; *industry* is provided with needed guarantees for investing in the development of products and infrastructure (e.g., in the case of Mobile Network Operators); *end-users* are more willing to onboard new technologies as the risk of being locked to a single vendor is (at least partially) mitigated; *policy makers* are able to take initiatives without the need to target specific solutions; *consumers* are guaranteed access to a variety of products and services. The overarching aim is to maintain open markets that encourage technology development and innovation. To this end, standardisation needs to be carefully balanced such that it does not favour or indirectly dictate product or vendor specific solutions.

A way to achieve this objective is by assuming an inclusive approach in standards development. This entails the engagement of multiple stakeholders with different scopes of authority and interest. We identify three coarse categories of SDOs, each one with its own distinct scope of work.

Standardization "within": The first category regards core technical SDOs strongly focusing on the technical aspects of the technology they study. The stakeholders involved in such SDOs primarily come from the industry and regard those who develop products or provide services. In such SDOs national/regional policymakers are also involved as technical solutions often require harmonising corresponding legislation (such as in the case of spectrum allocation). An example of such an SDO is the European Telecommunications Standards Institute (ETSI)[1] that has developed widely adopted telecommunications standards, such as the GSM.

"Outward" Standardization: The second category of SDOs regards harmonization of standards across different SDOs and regions. The aim is to harmonise the development of standards in the context of a globalised market and tightly interconnected world. An example of such an SDO is 3GPP,[2] a consortium with seven national or regional telecommunication standards organizations as primary members ("organizational partners") and a variety of other organizations as associate members ("market representation partners"). Example standards developed and maintained by 3GPP include the various generations of mobile telecommunications (3G, 4G, 5G, etc.).

---

[1] https://www.etsi.org/.
[2] https://www.3gpp.org/.

Standardization "beyond": The third category of SDOs regards high-profile international organisations where national and regional policymakers and regulators are actively involved alongside other stakeholders such as from industry, academia, and other. An example of such an SDO is the International Telecommunications Union (ITU)[3] whose activities are mainly driven by national delegations. Such SDOs have the mandate to develop policies at the international level and are crucial in the global management of resources, e.g., the ITU Radiocommunication Sector (ITU-R) plays a vital role in the global management of the radio-frequency spectrum and satellite orbits. At the same time, organizations of this category provide an inclusive forum for activities in emerging domains with the participation of a wider set of stakeholders. For example, Study Group 20 of ITU-T is very active in producing outputs (Recommendations, Supplements and Technical Reports) in the domain of Smart Cities and Communities.

1.3 Scoping the Future: Experiences in Standardisation for Smart Cities

Smart Cities employ digital technologies to improve processes and services with the aim of providing their citizens with improved quality of life and prosperity. In this context, industrial stakeholders have been considered technology providers and enablers, whereas local authorities and citizens have typically been regarded as end users and beneficiaries. However, recent technological advances and newly emerged ICT paradigms now provision more central roles for local authorities and citizens.

An indicative example is that of Neutral Hosting (NH) in Smart Cities. Neutral Hosting is a network model where telecommunications infrastructure is shared among mobile network operators (MNOs) for a pervasive infrastructure deployment (Paolino et al. 2019). The model allows to mitigate business issues related to thin client base in certain areas (e.g., rural areas) that have a detrimental impact on the ROI (return on investment) of deploying networking infrastructure (e.g., cellular masts). The NH model now provides the technical means for local authorities to deploy their own "dark" cellular infrastructure which can then lease to MNOs. This is also relevant with regard to mmWave infrastructure, where the local knowledge of the deployment area is of particular importance.

A second example is the paradigm of crowdsourced systems, which provisions the use of crowdsourcing methods in developing digital networked systems. In particular, the paradigm provisions the use of devices and infrastructure provided by the general public in the context of systems delivering a service. In academic literature the paradigm of crowdsourced systems has attracted attention, such as in developing distributed localised cloud services (Hosseini et al. 2019) and studying incentive mechanisms to encourage public engagement (Angelopoulos et al. 2015). However, being a novel and emerging

---

[3] https://www.itu.int/en/.

paradigm meant that key concepts and terms remained largely undefined, thus hindering the development of a commonly understood framework.

1.3.1 ITU-T Y.4205: Requirements and Reference Model of IoT-Related Crowdsourced Systems

Identifying the increasing interest showed by the academic community in the emerging paradigm of crowdsourced systems and its potential for future impact, a proposal to establish a corresponding work item was put forward in ITU-T Study Group 20: Internet of things (IoT) and smart cities and communities (SC&C).[4] ITU-T SG20 was chosen due to its mandate *"to develop international standards (ITU-T Recommendations) providing commonly agreed guidance for implementing the Internet of Things (IoT) and its applications, as well as smart cities and communities"*. Question 5, which studies emerging digital technologies, terminology and definitions, was considered to be the standardisation forum best fit for developing definitions and a reference architecture for crowdsourced systems. As a result, a work item was initiated in 2016 with the aim of developing a corresponding ITU-T Recommendation.

ITU-T Recommendations are standards defining how telecommunication networks operate and interwork (ITU). These can be accessed through the links below. ITU-T Recs have non-mandatory status until they are adopted in national laws. The level of compliance is nonetheless high due to international applicability and the high quality guaranteed by ITU-T's secretariat, and members from the world's foremost information and communication technology (ICT) companies and global administrations (ITU).

Recommendation Y.4205: *"Requirements and reference model of IoT-related crowdsourced systems"* (ITU-T 2019) was approved for publication in 2019. The standard introduces the concept of crowdsourced systems, as well as the reference model of IoT-related crowdsourced systems for the support of Internet of things (IoT) applications and services to be provided via systems employing crowdsourcing principles. It addresses IoT-related crowdsourced systems in terms of functional requirements and the reference model as well as identifying relevant security, privacy and trust issues (Fig. 1.1).

More specifically, *crowdsourcing* is defined as the practice of obtaining needed services, ideas, content or other system resources by soliciting contributions from a large, open and potentially undefined group of people, rather than from employees, suppliers or identified experts through an online open call by providing incentives (financial, social, or entertainment) to all or a subset of those crowd members who participate in the crowdsourcing activity (ITU-T 2019). Also, a *crowdsourced system* is defined as a system that employs crowdsourcing to augment their constituent infrastructure and the set of provided services or collected information. These are the first formal definitions in literature for these terms.

---

[4] https://www.itu.int/en/ITU-T/studygroups/2022-2024/20/Pages/default.aspx.

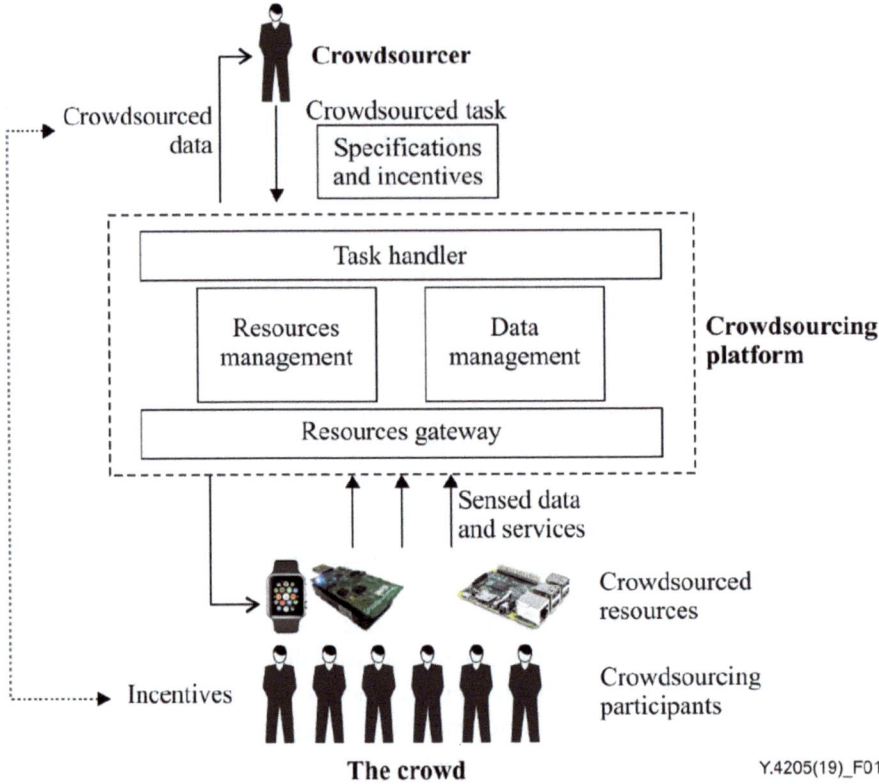

Fig. 1.1 The reference model for IoT-related crowdsourced systems (ITU-T 2019)

1.3.2 Lessons Learned and Experience Gained

The path towards successfully developing an international standard is an onerous but a rewarding one. First and foremost, the proponent should be familiar with the processes and various approval stages of the corresponding SDO. For newcomers, this may involve a steep learning curve, particularly if there is limited experience in standardisation in general. Support or collaboration with experienced colleagues that will double as mentors can be crucial in accelerating the process. Secondly, one should bear in mind that in the corpus of an SDO, almost by definition, all participants carry very high expertise. Albeit this expertise may be in different, not strictly technical topical areas—e.g., experts in patents with legal background—in the context of standardisation, this is highly relevant. Therefore, a proponent should be prepared to address any issues arising from language gap in the understanding of novel technical or scientific terms. Thirdly, in several cases agreement on publishing a standard is reached by consent (such as in the case of ITU-T) and therefore one should be ready to make compromises and reach the middle ground. As an example, in the development of Y.4205 specific algorithmic solutions were proposed for providing incentive mechanisms to crowdsources (the contributors from the public). However, the working group of experts raised concerns regarding whether making a reference to specific algorithms could prohibit future approaches by being considered as non-compliant. Overall, the experience gained from the development of Y.4205 has been a largely positive one. Although one needs to navigate a complex landscape of regulations, policies and balancing of opposing views, the process guarantees a high-quality outcome the broadest possible acceptance.

1.4 Conclusions

Smart Cities is a paradigm provisioning the use of digital services and networks to improve the quality of life and prosperity of citizens. The use of emerging ICT now supports the collection, curation, and processing of dense data in both space and time domains. This enables informed decision making and services planning in Smart Cities operations, thus driving further improvement and growth. The development of corresponding international standards is a key vehicle in developing solutions that are interoperable providing stakeholders with a common reference framework that guarantees an open market, a diverse set of available solutions and affordable costs.

The standardisation landscape is itself very diverse with a variety of standards development organisations with varying scopes and aims. They form a non-normative hierarchy with SDO activities "within", "outward", and "beyond" the scope of individual topical areas. The book you hold in hand in its following chapters provides valuable insights in standardisation activities in the domain of Smart Cities by high-profile SDOs across different regions of the world. This is a valuable collection of case studies providing the reader with a unique viewpoint that will inform the strategy of future standardisation efforts.

References

Angelopoulos C, Nikoletseas S, Raptis T, Rolim J1 (2015). Design and evaluation of characteristic incentive mechanisms in mobile crowdsensing systems. Simul Model Pract Theor. Elsevier

Eurostat (2024) Urban-rural Europe—economic activity in capital cities and metropolitan regions. Eurostat. Retrieved from https://ec.europa.eu/eurostat/statistics-explained/index.php?title=Urban-rural_Europe_-_economic_activity_in_capital_cities_and_metropolitan_regions&oldid=625445

Hosseini M, Angelopoulos C, Chai W, Kundig S (2019) Crowdcloud: a crowdsourced system for cloud infrastructure. J Cluster Comput. Elsevier

ITU (n.d.) ITU-T recommendations. Retrieved from https://www.itu.int/en/ITU-T/publications/Pages/recs.aspx

ITU-T (2019) Recommendation ITU-T Y.4205 requirements and reference model of IoT-related crowdsourced systems. Retrieved from https://www.itu.int/rec/T-REC-Y.4205/en

OECD (2015) Metropolitan areas. OECD Regional Statistics (database). Retrieved from https://doi.org/10.1787/data-00531-en

Paolino M, Carrozzo G, Betzler A, Colman-Meixner C, Khalili H, Siddiqui S, Simeonidou D, et al (2019) Compute and network virtualization at the edge for 5G smart cities neutral host infrastructures. In: IEEE 2nd 5G World Forum (5GWF). IEEE

Prof. Constantinos Marios Angelopoulos is Professor of Networked and Sensing Systems at Bournemouth University (U.K.), working on future and emerging paradigms of networks and networked systems. He is the founding Program Leader of the three Internet of Things Master courses and founded and leads the BU Open Innovation Lab. Since 2018, he serves at the International Telecommunication Union as Rapporteur in ITU-T SG20: Internet of things (IoT) and smart cities and communities (SC&C). His research has been published in highly esteemed peer-reviewed journals and conferences and has attracted external funding from several funding bodies (H2020, Horizon Europe, DCMS) and directly from Industry.

Dr. Ramy Ahmed Fathy is a senior standardisation expert carrying more than 15 years of experience in senior management roles. He is the Vice Chairman of ITU-T Study Group 20 on IoT and its Applications including Smart Cities and Communities, Co-Chairman of ITU-T SG20 Working Party 1, Chairman of Regional Group Africa, Chairman of ITU/FAO Focus Group on AI and IoT for Digital Agriculture. He is also Chief Expert of Digital Technologies and Public Policy in National Telecom Regulatory Authority, Egypt. His 15 years of key expertise include projects in the fields of cryptography (PKI), automation, analogue TV switch-off and DTV transformation, 4G business case development, market modelling, intelligent transportation systems (design—specifications—business case), and smart cities services ICT requirements and specifications.

Part I
The Context of Smart City Standards

What KPIs Can Reveal and Help Achieve

Barbara Kolm and Victoria Schmid

Abstract

Today, cities account for almost 50% of the world's population and 80% of the global GDP. These percentages are only expected to grow as more people move into the cities. Urbanization is a source of economic growth and development, but it can also be a source of challenges: overcrowding, disease, pollution, congestion, crime, and inequality (UN Trade and Development (UNCTAD), Four key challenges facing least developed countries. UN Trade and Development, 2022). If policy makers want to mitigate these risks, they have to engage in smart and sustainable city planning. The United for Smart Sustainable Cities (U4SSC), an initiative aimed at helping cities become more sustainable, can provide guidance to city planners who intend to make their cities resilient, sustainable, and inclusive.

Keywords

Smart sustainable city · KPIs · Sustainable development goals · U4SSC · ITU

B. Kolm (✉) · V. Schmid
Austrian Economics Center, Vienna, Austria
e-mail: b.kolm@austriancenter.com

V. Schmid
e-mail: v.schmid@austriancenter.com

B. Kolm
Hayek Institut, Vienna, Austria

2.1 Introduction

The phrase "smart city" inevitably recalls scenes from sci-fi movies: self-driving cars, virtual reality and robot assistants. These cities undoubtedly exist. Chinese and American cities alike are at the forefront of self-driving car testing and implementation. This is however a reduction of what a smart city is. A smart city can adopt technological innovations such as automated vehicles, multifunctional robots, or augmented reality. However, what makes a city "smart" goes beyond these technological amusements. Smartness does not embellish cities through technological innovations. But then, what makes a city smart? Robots? LED screens? To help cities become "smart", a set of initiatives such as the *United for Smart Sustainable Cities* initiative have been developed to guide cities in their path to "smartness" and to help them assess their progress towards it (International Telecommunication Union (ITU) n.d.).

Today, there are three internationally recognized bodies that produce guidelines or standards to help cities in their pursuits to become smart: the International Telecommunication Union (ITU), the International Organization for Standardization (ISO), and the International Electrotechnical Commission (IEC). These bodies produce standards within their expertise.

The International Telecommunication Union (ITU) is at the forefront of issuing guidelines for cities in their path towards "smartness" and sustainability. Established in 1865 to manage the first international telegraph networks and recognized in 1947 by the United Nations as its specialized agency for telecommunications, today the ITU is the United Nations specialized agency for information and communication technologies (ICTs) (International Telecommunication Union (ITU) (n.d.-b)). Among many other initiatives, the ITU launched in May 2016 the United for Smart Sustainable Cities (U4SSC) initiative (United Nations Economic Commission for Europe (UNECE) and International Telecommunication Union (ITU) n.d.). The U4SSC was launched to achieve the United Nations' Sustainable Development Goal (SDG) 11: Make cities and human settlements inclusive, safe, resilient, and sustainable. The U4SSC is a global public policy platform intended to promote the adoption and implementation of information and communication technologies and digital technologies to help cities transition into smart sustainable cities.

Today, cities account for almost 50% of the world population and 80% of the global GDP and are "responsible for close to 70% of global CO_2 emissions associated with energy consumption (International Telecommunication Union (ITU) 2016)." The importance of cities as economic and cultural hubs is undeniable and bound to increase as the share of the global population in cities is projected to rise from its current 48% to a staggering 58% in 2070 (UN-Habitat 2022). Population growth and urbanization will inevitably give rise to opportunities as well as to challenges. If unaddressed, urbanization, however, will pose significant challenges to citizens and governments alike. Without proper infrastructure, cities might be unable to deal with the economic, social, and environmental challenges that might arise because of population growth. Economic prosperity, social wellbeing, and environmental sustainability depend on policy makers' abilities to anticipate and address those future challenges. Smart and sustainable planning of cities is necessary to avert future crises.

2.2 Why Cities?

For the past 70 years, the world has witnessed notable population growth and urbanization rates. "In 1950, most people lived in rural areas, followed by towns and semi-dense areas, while cities were the least inhabited (UN-Habitat 2022)." Today, the order is reversed: most people live in cities, and the fewest people inhabit rural areas. By 2020, around 48% of the global population lived in cities, while 29% lived in towns and semi-dense areas, and 22% in rural areas. Population growth has decelerated and is expected to slow down even more. Urbanization, on the other hand, is expected to increase. According to the UN World Cities Report's estimates, the percentage of the world population inhabiting cities is expected to grow to 58% by 2070, while the world population in towns and semi-dense areas is expected to decrease to 24% (UN-Habitat 2022).

Urbanization is the inevitable consequence of a higher per-capita income in cities compared to that of rural areas. However, migration to urban areas is nothing new. The industrial revolution reshaped most of the world economy, turning rural agrarian societies into industrial economies; this, in turn, led to urbanization as labor shifted from farms to factories. Many laborers in the UK, the Netherlands and the US looking for better opportunities resettled in urban areas where wages were higher compared to those of the countryside. Nonetheless, even if wages were higher, working conditions in cities were dreadful. Death rates in cities were higher in comparison to those of rural areas:

> Urban mortality was indeed so high that, were it not for continual migration from the countryside, the cities would have faded from the earth. In London from 1580 to 1650, for example, there were only 0.87 births for every death. Without migration the city would have lost about half percent of its population every year. Early towns were generally crowded and unsanitary, so that infectious diseases such as plague, typhus, dysentery, and smallpox spread quickly (Clark 2010).

Despite London being one of the richest cities in the world, the life expectancy of its inhabitants at the end of the eighteenth century was merely 23 years—a lower life expectancy than that of most pre-industrial societies. It would take almost a hundred years for this to improve. As Jared Diamond noted: Not until the beginning of the twentieth century did Europe's urban populations finally become self-sustaining: before then, constant immigration of healthy peasants from the countryside was necessary to make up for the constant deaths of city dwellers from crowd diseases (Diamond 2017).

These unsanitary conditions have been eradicated in most Western cities. Today, London is no longer the unsanitary and death-ridden slum it was during the Georgian era.

According to the latest released data from the 2020–2022 Office of National Statistics census, life expectancy in London was 79.1 years as of the 2020–2022 period—almost 56 years longer than the average life expectancy of 23 in 18th-century England. Unsanitary conditions, however, are still prevalent in the Global South cities where overcrowding, poor sanitary conditions, crime, scarce resources, climate change, outdated infrastructure and pollution are rife. As of 2019, the 20 countries with the lowest life expectancy were in the Global South (Table 2.1). In 2019, these 20 countries had a meager life expectancy at birth of only 60.02 years.

The challenges that the Global North cities face are different from those of the Global South. However, there are some overlaps: overcrowding, rising crime rates, unemployment, climate change, outdated infrastructure, among many others. These and other challenges exclusive to the Global North such as an ageing population, social and political polarization, and a shrinking labor force jeopardize economic productivity and the welfare of the cities' inhabitants.

Table 2.1 Average life expectancy in the countries with the lowest life expectancy at birth (World Health Organization (WHO) 2020)

	Country	Life expectancy at birth (in years)
1	Lesotho	50.7
2	Central African Republic	53.1
3	Somalia	56.5
4	Eswatini	57.7
5	Mozambique	58.1
6	Kiribati	59.4
7	Chad	59.6
8	Guinea-Bissau	60.2
9	Zimbabwe	60.7
10	Sierra Leone	60.8
11	Guinea	61
12	Botswana	62.2
13	Equatorial Guinea	62.2
14	Cameroon	62.4
15	Democratic Republic of the Congo	62.4
16	Zambia	62.5
17	Nigeria	62.6
18	Burkina Faso	62.7
19	Mali	62.8
20	South Sudan	62.8

Urbanization will pose significant challenges in the future. Economic growth and citizen flourishment depend on policy makers' abilities to deal arising issues. But to tackle these issues effectively policy makers will have to anticipate change, correct the course of action and implement effective policies.

2.3 What is a Smart City?

Today, cities account for almost 50% of the world population and 80% of the global GDP. In both the Global North and South, citizens leave their communities in search of better opportunities in urban areas. This is hardly any surprise. Higher population densities, lack of opportunities in rural areas and centralization make cities desirable destinations for those looking to make a living. But that is only half of the story: cities mean opportunities for individuals while for countries, "cities are the engines of productivity and the workhorses of development (Fikri and Juni Zhu 2015)." Cities themselves act as economic and cultural networks that help connect workers with firms, craftsmen with customers and artists with an audience.

Cities thrive on agglomeration. A higher population density or agglomeration creates networks, brings in new stakeholders and accelerates processes that are crucial for development (Lozano-Gracia and Soppelsa 2019). This, however, gives rise to new challenges: disease, pollution, congestion, crime and inequality. It is partly inevitable. Because the same crowds that allow ideas to spread allow diseases to spread. The networks that connect people and allow markets to flourish allow criminal networks to operate. The closeness that facilitates trade facilitates crime, too. Competitiveness gives rise to congestion and congestion to pollution. And the cities' wealth attracts poverty as the destitute find work and opportunities in the cities, hence creating inequality. This paradoxical phenomenon is labelled the "demons of density"—a concept alluded to by Harvard economist Edward Glaeser in his book *The Triumph of the City* (Glaeser 2008).

As urbanization is poised to increase, so are the density demons. These demons ought to be cast out and if policy makers want to do so effectively, they must engage in sustainable and smart city planning. And the United 4 Smart Sustainable Cities (U4SSC) initiative provides the guidelines and tools to do so.

But first, what is a "smart sustainable city"? The ITU, the UN telecommunication agency, defines a smart sustainable city as:

> An innovative city that uses information and communication technologies (ICTs) and other means to improve quality of life, efficiency of urban operation and services, and competitiveness, while ensuring that it meets the needs of present and future generations with respect to economic, social, environmental as well as cultural aspects (International Telecommunication Union (ITU) 2016).

Succinctly, it is a city that uses technology and data to make better decisions and consequently to improve the quality of life of its citizens and future generations. Future challenges warrant policy makers to design and implement strategies capable of mitigating urbanization problems.

The path towards "smartness" and sustainability can be difficult. Without the appropriate guidelines, knowledge, expertise and support, governments can easily misdirect their efforts to turn their cities into smart sustainable cities, wasting their citizens' resources in the process. For this reason, different initiatives such as the United 4 Smart Sustainable Cities (U4SSC) initiative have been developed to help cities in their pursuit of becoming smart and sustainable. The U4SSC is a novel initiative currently supported by 19 UN agencies and programs to achieve Sustainable Development Goal 11: "Make cities and human settlements inclusive, safe, resilient and sustainable".

The ITU defines the U4SSC as a "global smart sustainable city initiative which provides an international platform for information exchange, knowledge sharing and partnership building, with the aim of formulating strategic guidance to achieve the Sustainable Development Goals (SDGs) and implement the New Urban Agenda and other international agreements (International Telecommunication Union (ITU) 2021)."

Its objectives are twofold:

- To generate guidelines, policies, and frameworks for the integration of ICTs into urban operations, based on the SDGs, international standards and urban key performance indicators (KPIs).
- To help streamline smart sustainable city action plans and establish best practices with feasible targets that urban development stakeholders are encouraged to meet. (ITU—Report: United 4 Smart Sustainable Cities)

2.4 United for Smart and Sustainable Cities (U4SSC)

To hedge against future risks, city planners have already set their cities on the path to "smartness" and sustainability. However, smartness and sustainability are very broad and nebulous objectives. How does a city become smart and sustainable? When has a city achieved smartness and sustainability? These questions might be difficult to answer without proper guidance. Bodies such as the ITU have sought to specialize in guiding cities and help them answer these questions.

Having helped over 150 cities and endorsed by the UN, the ITU´s U4SSC initiative is at the forefront of driving cities towards smartness and sustainability. U4SSC has developed two strategic initiatives: USSC hubs and KPIs for Smart Sustainable Cities.

- The U4SSC hubs are platforms designed to sponsor cooperation between the public and private sector (start-ups, multinationals, research institutions, etc.) to "facilitate

the digital transformation in cities and communities, while enabling technology and knowledge transfer (International Telecommunication Union (ITU) (n.d.-c))." Today two country hubs have been set up. The first one in Vienna, Austria and the second one Kyebi, Ghana. Both Hubs are committed to bring together government, industry and civic leaders to make cities safer, more resilient and sustainable.
- Key Performance Indicators (KPIs) are guidelines intended to help cities in their path towards sustainability and development. Additionally, they are self-assessment tools that support "cities and communities worldwide in evaluating their level of smartness and sustainability (International Telecommunication Union (ITU) (n.d.-c))."

Both hubs and KPIs are integral to the U4SSC's mission. While the U4SSC hubs foster cooperation, the KPIs act as guidelines and assessment tools. KPIs, however, have a preeminent role; they are the core of the U4SSC initiative.

2.5 The Participation Process

A city interested in participating in the U4SSC KPIs project should submit a letter to the ITU-U4SSC Secretariat, the body in charge of providing process guidance to cities during duration of the U4SSC KPIs project, conveying its interest in participating in the project.

The letter should convey the city's interest in participating in the project and its willingness to undertake all the necessary activities required by the U4SSC KPIs project. The city pledges to collect all the necessary data required to perform the relevant analysis, to submit the data to the ITU-U4SSC Secretariat, and to allow the data to undergo a verification process by a U4SSC approved verifier. Additionally, the letter should describe the city's efforts performed so far and the city's vision of becoming a smart sustainable city. (A letter template to request participation in the U4SSC KPIs project can be found in Annex 2 of the *Key performance indicators: A key element for cities wishing to achieve the Sustainable Development Goals* document.)

A city participating in the U4SSC KPIs project pledges to abide to the applicable Terms and Conditions of the KPI project. The Terms and Conditions describe the responsibilities of the Participating City, the responsibilities assumed by the ITU U4SSC Secretariat and the associated costs of the project.

The responsibilities of the participating city include but are not limited to:

- Gathering the data in a 3-to-6-month time span. The assistance of a qualified third-party consultant is permitted.
- Collecting and recording the required information in an Excel template provided by the ITU-U4SSC Secretariat
- Paying for the travel and accommodation expenses of an independent verifier who will conduct a series of interviews and will produce a final Verification Report for the

city. The verifier shall be designated by the ITU-U4SSC Secretariat. Verification can also be performed remotely, in which case the city will not have to pay for travel and accommodation expenses, only for the verifier fees.
- Submitting the data on a date agreed by the city, ITU and the verifier.
- Paying for the costs of the whole project.

The responsibilities assumed by the ITU U4SSC Secretariat include but are not limited to:

- Ensuring that each participating city is well informed about the processes conducted during the project.
- Appointing a competent and accomplished verifier who can carry out the verification process.
- Deciding on a suitable UN body staff member who can supervise the verification work carried out by the verifier.

(More information about the Terms and Conditions can be found on Annex #1 of the *Key performance indicators: A key element for cities wishing to achieve the Sustainable Development Goals* document.)

2.6 KPIs

KPIs are metrics that track an organization's performance towards a goal or objective. First, objectives are set. Then, a set of related KPIs are defined to measure the progress achieved towards that goal. The U4SSC KPIs are no different. They are metrics that serve as guidelines as well as assessment tools. They are guidelines as they point towards objectives to which cities should focus to become smart and sustainable. And they are assessment tools as they help monitor performance.

The U4SSC initiative has three objectives: help cities become smarter, help cities become more sustainable and help cities achieve the UN Sustainable Development Goals (SDGs). The initiative has therefore devised 91 KPIs to assist cities in achieving these three goals. By applying these KPIs, cities can evaluate their progress. They can also evaluate the role that ICTs and digital technologies play in bringing them closer to the three objectives. The KPI indicators are reliable and comparable as they provide a "consistent and standardized method to collect and report the data needed to quantify, measure, report and monitor performance and progress (International Telecommunication Union (ITU) (n.d.-c))."

Each KPI falls under three dimensions: Economy, Environment or Society and Culture. Dimensions are further broken down into sub-dimensions, as well as categories. In total, there are three dimensions, seven sub-dimensions and twenty-seven categories. The following table (Fig. 2.1) shows a sample arrangement of some KPIs on the Society and Culture dimension. The Adult Literacy KPI (the fourth KPI listed on the table) is part of the Society and Culture dimension, of the Education, Health and Culture sub-dimension and part of the Education category. Furthermore, it is a core type KPI.

Figures 2.2, 2.3 and 2.4 are three diagrams that graphically show the complete breakdown into dimensions, sub-dimensions, and categories. (To see the breakdown of the 91 KPIs refer to the USSC: Collection Methodology for Key Performance Indicators for Smart Sustainable Cities). Dimensions focus on broad categories, while sub-dimensions and categories focus on more specific areas of performance and progress. KPIs are further categorized as core or advanced indicators.

> Core indicators are those that should be able to be reported on by all cities, provide a basic outline of smartness and sustainability and higher levels of performance can generally be achievable. Advanced indicators provide a more in-depth view of a city and measure progress on more advanced initiatives; however, they may be beyond the current capabilities of some cities to report or implement (International Telecommunication Union (ITU) 2017).

There are 91 KPIs. Within each KPI there is an explanation indicating the rationale underpinning the KPI, the methodology used to calculate the KPI, the KPI unit of measurement, the data sources or relevant databases from where information could be obtained and the Social Development Goal (SDG) references. The following table (Fig. 2.5) is an example of a KPI: the School Enrollment KPI.

This is an example of a KPI. This KPI belongs to the Society and Culture dimension, to the Education, Health, and Culture sub-dimension, and to the Education Category. The Rationale/Interpretation/Benchmarking row substantiates the relevance of this indicator, provides a brief explanation of the indicator, of how to interpret it, and the range within which desirable values are to be found. The Methodology row describes the process to calculate the indicator. The Unit row refers to the unit of measurement to be used in this KPI. The Data Sources/Relevant Databases row alludes to sources where the data could be found. The SDG References refer to the UN Sustainable Development Goal that it aims to address.

There is one of these tables for each of the 91 KPIs. Tables can be found in the *USSC: Collection Methodology for Key Performance Indicators for Smart Sustainable Cities* document.

Indicators have been carefully thought up to ensure that they are relevant in measuring a city's progress in becoming smarter and more sustainable. The standardized nature of the KPIs and the verification process by the ITU ensure that indicators can be comparable throughout cities (International Telecommunication Union (ITU) 2017).

Collecting the data, calculating the KPIs and performing the necessary analyses allows a city to self-assess its performance and the achieved progress to smartness and sustainability.

Dimension	Sub-Dimension	Category	KPI	Type	Type
Society and Culture	Education, Health and Culture	Education	Student ICT Access	Core	SMART
			School Enrollment	Core	STRUCTURAL
			Higher Education Degrees	Core	STRUCTURAL
			Adult Literacy	Core	STRUCTURAL
		Health	Electronic Health Records	Advanced	SMART
			Life Expectancy	Core	STRUCTURAL
			Maternal Mortality Rate	Core	STRUCTURAL
			Physicians	Core	STRUCTURAL
			In-Patient Hospital Beds	Advanced	STRUCTURAL
			Health Insurance / Public Health Coverage	Advanced	STRUCTURAL
		Culture	Cultural Expenditure	Core	STRUCTURAL
			Cultural Infrastructure	Advanced	STRUCTURAL
	Safety, Housing and Social Inclusion	Housing	Informal Settlements	Core	STRUCTURAL
			Housing Expenditure	Advanced	STRUCTURAL
		Social inclusion	Gender Income Equity	Core	STRUCTURAL
			Gini Coefficient	Core	STRUCTURAL
			Poverty	Core	STRUCTURAL
			Voter Participation	Core	STRUCTURAL
			Child Care Availability	Advanced	STRUCTURAL
		Safety	Natural Disaster Related Deaths	Core	SUSTAINABLE

Fig. 2.1 KPI society and culture with sub-dimensions and categories (International Telecommunication Union (ITU) 2017)

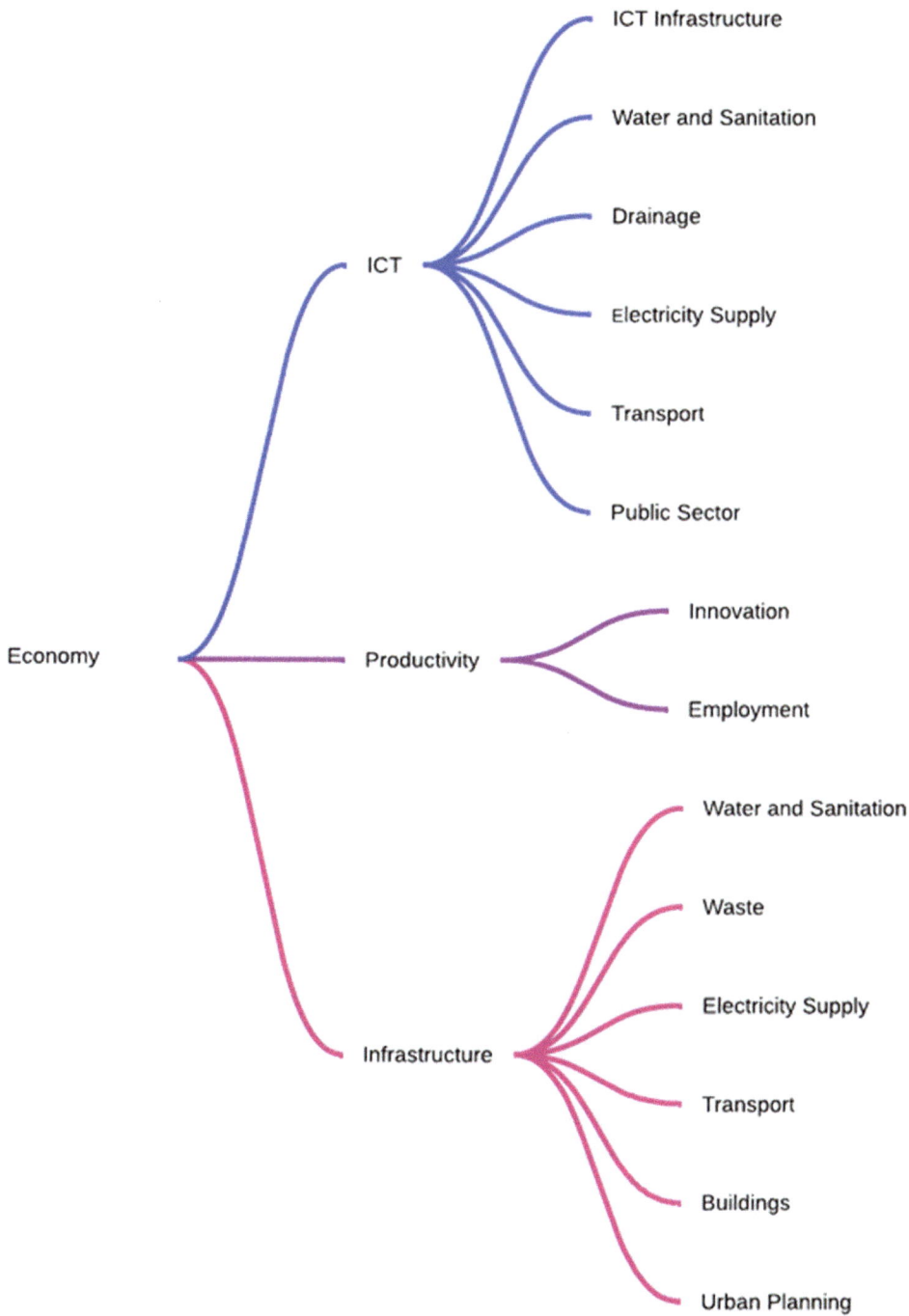

Fig. 2.2 Breakdown into sub-dimensions and categories of the economy dimension of the KPIs project (International Telecommunication Union (ITU) 2017)

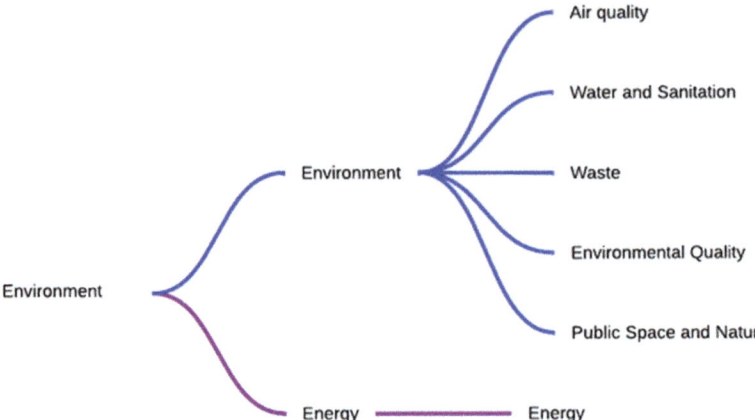

Fig. 2.3 Breakdown into sub-dimensions and categories of the environment dimension of the KPIs project (International Telecommunication Union (ITU) 2017)

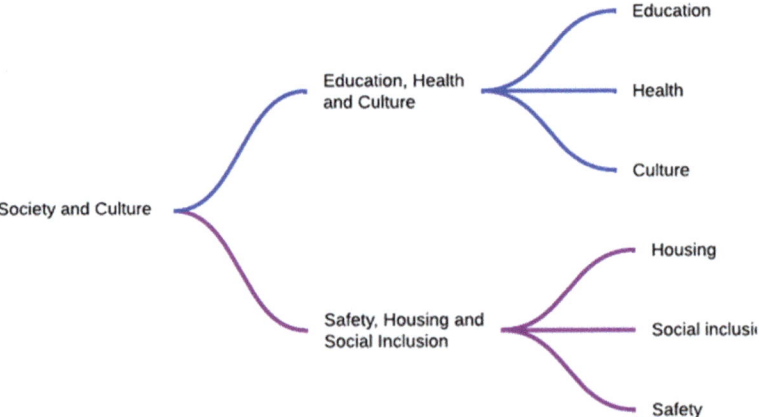

Fig. 2.4 Breakdown into sub-dimensions and categories of the society and culture dimension of the KPIs project (International Telecommunication Union (ITU) 2017)

However, self-assessment is just a part of the project, feedback is another relevant part of it. Cities receive feedback once they have submitted the data to the independent verifier.

Dimension	Society and Culture				
Sub-Dimension	Education, Health and Culture				
Category	Education				
KPI Name	School Enrollment				
KPI No.	SC: EH: ED:2C	Type:	Core	Type:	Structural
Definition / Description	Percentage of school-aged population enrolled in schools				
Rationale / Interpretation / Benchmarking	Education is essential to human development. It is also an indicator of the future potential of a city, its inhabitants and work force. A city should report on public and private enrolment as well as recognized religious and home schools that meet defined governmental standards. An improving trend and higher values are considered positive.				
Methodology	Calculate as: Numerator: Number of students in primary and secondary levels in public and private schools. Denominator: Total number of the school aged population. Multiply by 100				
Unit	Percentage				
Data Sources / Relevant Databases	Enrolment data can be collected from local school boards / authorities or regional / national education departments.				
SDG Reference(s)	SDG Target 4.1: By 2030, ensure that all girls and boys complete free, equitable and quality primary and secondary education leading to relevant and effective learning outcomes.				

Fig. 2.5 KPI school enrollment (International Telecommunication Union (ITU) 2017)

2.7 The Verification Process

The first step for a city to get involved in the U4SSC KPIs project is to pay the project fees. Once these have been paid, the city will be accepted and welcomed into the U4SSC initiative. After the payment, the ITU will, in turn, assign an independent third party who is certified as a U4SSC Key Performance Indicators for Smart Sustainable Cities Verifier to conduct a verification of the data submitted by the city that has signed up for the U4SSC initiative. The verifier ensures that the submitted data are complete, accurate, supported, and meets the requirements found in the *U4SSC Collection Methodology for Key Performance Indicators for Smart Sustainable Cities* document.

Once the ITU, the verifier, and the city are in contact, the verifier will provide the city with the U4SSC data collection template. This document indicates all the KPIs that are to be collected, reported, and assessed. The city will, in turn, proceed to gather and collect all the relevant data and calculate all the KPIs required for the assessment, in accordance with the *U4SSC Collection Methodology for Key Performance Indicators for Smart Sustainable Cities*. Once the city has gathered all the relevant information, it will proceed to fill in the

U4SSC data collection template with the collected data and information received from the various entities in their city. (U4SSC Collection Methodology for Key Performance Indicators for Smart Sustainable Cities) The city will make the data and information available to the verifier with all its supporting documentation (both paper and electronic) and data sources. If the verifier has any additional questions, the city will provide complete and timely explanations to the verifier's questions.

Once the data and sources have been made available, the verifier will start with the verification process of the city's U4SSC KPIs data. The verifier will ensure that the submitted data meets the requirements of the U4SSC KPIs standards.

Verification can be performed onsite or remotely. An onsite verification will require the city to pay for the agreed verification costs plus any related travel expenses (e.g., flight ticket, accommodation and per diem costs). Remote verification will only require the city to pay for the agreed verification costs. An onsite verification usually takes place at the city's offices and has the advantage that supporting documents can be viewed in-person and questions can be addressed quickly, minimizing delays in completing the verification. A remote verification will take place electronically. In that case, the verifier asks the city to provide any supporting documents required for the verification electronically. (ITU's Implementation of the U4SSC KPIs: Verification Process.)

The verifier will proceed to examine the city's "reports, books, records, documents, and other information (International Telecommunication Union (ITU) 2017)" (collectively referred to as 'supporting documents such as annual report, financial filings, policy documents, etc.). If any questions arise during the verification period, the city will have to promptly answer and discuss these with the verifier. The city can also raise any questions or concerns with the verifier at any time. Good data collection and reporting, cooperation and prompt responses will optimize the verification process. Lack of compliance with the verifier's request will result in delays.

If the verifier has found that the city's data is complete, accurate, supported through documentation, and in compliance with the U4SSC KPIs standards, then the verification process will be regarded as complete. The city will then receive a notification of completion, and the verification will be closed. Otherwise, the verifier will have to perform a re-verification. In this case, the city will receive a notification substantiating the re-verification. The verifier will list additional documentation that is required to complete the re-verification. In the event of disagreement regarding the re-verification, the city can contact the verifier to provide any other documents that support its position and settle the disagreement. The verifier will carefully consider any explanations and respond to the city's questions, copying the ITU/U4SSC Secretariat. If it remains unresolved, the city can contact the ITU/U4SSC Secretariat to discuss the matter. If the issues arise due to the city not having the original supporting documents, "it can provide reasonable alternatives or a combination of documents that also support the data being reported (International Telecommunication Union (ITU) 2017)." "If the city cannot provide any documents or sources to support the data it is reporting, it can discuss this with the verifier who will work with the city to find ways to confirm the reported data (International Telecommunication Union (ITU) 2017)." (ITU's Implementation of the U4SSC KPIs: Verification Process.)

If the verification has been successful, a notification will be sent to the city, and no adjustments will be made to the city's data from that point onward. The city will also receive an attestation from ITU upon completion of its U4SSC KPIs project, "stating that it was successfully evaluated based on a UN standard (International Telecommunication Union (ITU) 2017)." Finally, the verifier will deliver a City Snapshot and a Verification Report. These two reports provide feedback on the city's implementation of the KPIs.

The Verification Report is a comprehensive report on the city's implementation of the SSC KPIs. The Verification Report highlights the KPIs that the city has successfully reported and provides results of the performance attained by the city in each of these indicators. In addition to providing relevant feedback, the Verification Report provides high-level policy recommendations and suggestions for improvement to further achieve smartness and sustainability. (ITU's Implementation of the U4SSC KPIs: Verification Process.)

In addition to the Verification Report, the verifier produces a City Snapshot. This is a visual overview of "city's performance to benchmark in the Key Performance Indicators for Smart Sustainable Cities (International Telecommunication Union (ITU) 2017)." While the Verification Report provides individual feedback on each of the 91 KPIs feedback on all of the KPI's, the City Snapshot highlights performance only at the dimension and subdimension level.

Verification reports and City Snapshots are included as feedback as part of the U4SSC KPIs project. However, in addition to those reports, a city can request the ITU-U4SSC Secretariat's assistance to develop a City Factsheet or Case Study. These are granular reports which communicate in detail the outcomes and achievements of the city after implementing the U4SSC KPIs. These reports describe the city's smartness and sustainability initiatives and illustrate how the city has benefited from their implementation, bringing it closer to achieving the Social Development Goals (SDGs). Both the City Factsheet and the Case Study are tools that scrutinize a city's policies and report the city's progress towards achieving its sustainability and smartness goals.

The City Factsheet contains a comprehensive analysis on the performance of a city in each of the Key Performance Indicators for Smart Sustainable Cities. It documents the initiatives and activities that the city has undertaken to become smarter and more sustainable and links these activities to the KPIs. The factsheet then analyses the effectiveness of the city's activities. "Given that the KPIs for SSC are connected to the targets of the Sustainable Development Goals, the factsheet also provides a general overview on the city's progress towards the SDGs (International Telecommunication Union n.d.-d)." Additionally, the factsheet provides tailored policy recommendations to address the city's shortcomings in KPI performance.

Case studies are tailored reports recounting a city's experience in transforming into a smart sustainable city. These reports review the programs implemented by a city in becoming an SSC, the activities undertaken as part of the KPI program and the city's progress in reaching the objectives of its smart strategies and meeting the Sustainable Development Goals. Results of Case Studies provide valuable insights into other cities' experiences in becoming smart and sustainable and provide valuable lessons to other cities that intend to become SSCs. Additionally, the reports provide the major findings arising from the implementation of the U4SSC Key Performance Indicators (KPIs).

If correctly implemented the KPI methodology allows a city to asses and consequently to understand its strengths and shortcomings. Through the KPI methodology, a city can set new strategic goals, perform longer term trend analysis and assess changes in the values KPIs over time.

2.8 The City of Wels, Austria

Wels was the first city in Austria to undergo the audition. The city government wanted a thorough analysis of its potential and future development opportunities. With this detailed information the city officials would then be able to plan strategies compatible with the specific goals that correspond to the UN's Sustainable Development Goals (SDGs) and the requirements of the EU's European Green Deal.

In 2020, Wels applied for participation in the U4SSC initiative. The analysis was started in the second quarter of 2020 and then not completed until the third quarter of 2020 due to the COVID-19 pandemic. It is based on figures from 2018 and 2019, which is why the effects of the coronavirus crisis are not included.

The results show the extent to which Wels has adopted information and communication technologies (ICT) in its public infrastructure and services. The city scores particularly well in the economy. There is potential for investment, sustainability improvement and integration of ICT in areas such as public transport, research and development activities, employment indicators and urban planning.

Wels performs similarly well in the environment, with most of the indicators for air quality, water quality, environmental quality, green spaces, waste management and energy meeting sustainability thresholds. Finally, most of the social and cultural indicators related to education, health, culture, housing and safety are also in the proverbial green zone.

The results for Wels are documented in the U4SSC City Snapshot: Wels, Austria (November 2020). Figure 2.6, included in the City Snapshot, shows the results attained by the city of Wels at the subdimension level. This is a graphical representation of performance. This is the first in a series of U4SSC reports that will evaluate and progressively develop the results of the implementation of the U4SSC Key Performance Indicators (KPIs). These reports (verification report, factsheet and case study) will subsequently provide useful analysis, actionable recommendations and other important insights for the future development of the Smart Sustainable City Wels.

The results were presented at a press conference, where Dr. Cristina Bueti from the UNO-Organization ITU Said,

the city of Wels was a great candidate for implementing the U4SSC KPIs for several reasons. This historic city is one of the most important commercial and cultural centers in Austria. Its beautiful environment and richness in culture have attracted tourists and visitors to the city from all over the world. The city's strong foundation for tourism is complemented by a diverse economy which comprises of a prominent service sector as well as industrial and agricultural

Fig. 2.6 Performance of the city of Wels, Austria (United For Smart Sustainable Cities (U4SSC) 2020)

sectors. This dynamism has created exceptional synergy that has enabled Wels to drive innovation and transformation. The U4SSC KPIs can help the city to refine its innovation strategy and align it with important global commitments such as the Sustainable Development Goals. That is why Wels was the ideal city to implement the U4SSC KPIs (Wels Verwaltung 2021).

2.9 Conclusions

Today, almost half of the world's population lives in cities. Cities account for 80% of the global GDP and are 'responsible for close to 70% of global CO_2 emissions associated with energy consumption (International Telecommunication Union (ITU) 2016). In both developed and undeveloped countries, cities remain alluring as ever as they provide a clear path to prosperity. Urbanization is a source of economic growth and development, but it can also be a source of challenges: overcrowding, disease, pollution, congestion, crime and inequality (UN Trade and Development (UNCTAD) 2022).

The future poses relevant challenges to cities in both the developed and undeveloped world. While they both share some common challenges such as overcrowding; rising crime rates; unemployment; climate change; outdated infrastructure, they face distinct challenges, too (UN Trade and Development (UNCTAD) 2022). On the undeveloped world, water, air, and soil pollution, along with energy poverty and climate vulnerability, are some of the most pressing issues. While on the developed world, population ageing, social and political polarization, and a shrinkage of the labor workforce are some of the most pressing issues. These challenges are expected to exacerbate as urbanization is projected to grow in the coming decades. If governments want to mitigate contingencies, they will have to anticipate future risks, correct the course of action and engage in smart city planning. Building cities that work—with access to clean water, sanitation, functional waste disposal, efficient transportation, safe and healthy spaces—requires intensive policy coordination and investment choices (World Bank 2014).

Technology has become the backbone of modern life. Policymakers can use the tools available to make cities smarter and more sustainable. A smart city is simply one that uses technology and data to make better decisions and consequently to improve the quality of life of its citizens and future generations. Future challenges warrant policy makers to design and implement strategies capable of mitigating urbanization problems.

The U4SSC initiative can provide guidance to city planners who intend to make their cities resilient, sustainable, and inclusive. The U4SSC's KPI project gives cities the necessary tools and guidance to undertake projects that will make cities smarter and more sustainable. Additionally, the project will aid cities self-assess their progress towards these two goals. Participating cities receive two reports, the City Snapshot and a Verification Report. These two reports provide feedback on the city's implementation of the KPIs. The Verification Report is a comprehensive report on the city's implementation of the SSC KPIs. The City Snapshot is a visual overview of it. In addition to these reports, cities can request additional reports from the ITU-U4SSC Secretariat such as Factsheets or Case Studies. These are granular reports that provide a detailed account of the outcomes and achievements of a city after implementing the U4SSC KPIs. Case Studies provide valuable insights into cities' experiences in becoming smart and sustainable and can be used by other cities and policymakers to draw conclusions about the correct policy implementation. Even if cities are different and face different challenges, drawing from other cities' experiences can help policymakers better mitigate and overcome certain risks.

"Building cities that work requires intensive policy coordination and investment choices (World Bank 2014)." The KPI project is one of the most comprehensive assessment tools to evaluate a city's performance in becoming sustainable, inclusive, and resilient. If correctly implemented, the KPI project will allow policymakers to assess the effectiveness of the policies they have implemented and to derive relevant insights from policy recommendations. Economic growth and human flourishing is dependent on policy makers' abilities to anticipate risks and to implement relevant policies. Without the proper policies, cities might be unable to deal with the economic, social, and environmental challenges that might arise as a consequence of population growth. Economic prosperity, social wellbeing, and environmental sustainability depend on policy makers' abilities to anticipate and address those future challenges.

2.10 Revision Questions with Answers

(1) What is a smart sustainable city? Answer with your own words

A smart sustainable city is one that uses technology and data to make better decisions and consequently to improve the quality of life of its citizens and future generations.

(2) Why is smart city planning more relevant than ever?

Smart city planning is more relevant than ever, as the percentage of the world population living in cities is poised to increase to 58% by 2070. Urbanization will bring relevant challenges that policy makers have to take today into consideration.

(3) What are the two strategic initiatives of the U4SSC? Provide a brief explanation.
- The U4SSC hubs are collaboration platforms designed to bring public and private sector together with the purpose of facilitating the digital transformation in cities.
- KPIs are guidelines designed to advise cities on how to achieve smartness and sustainability and help them self-assess their progress towards it.

(4) How do KPIs work?

KPIs work as guidelines as they point towards objectives to which cities should focus to become smart and sustainable, and they work as assessment tools as they help cities monitor performance.

(5) What are the default final products of the verification conducted by the independent verifier? Briefly describe them.

As a default, the verifier will provide a Verification Report and a City Snapshot. A Verification Report will describe the performance by the city in each of these KPIs. Within the report some policy recommendations will be provided to the city. The City Snapshot is a visual overview that highlights performance only at the dimension and subdimension level.

References

Clark G (2010) A farewell to alms: a brief economic history of the world. Princeton University Press

Diamond JM (2017) Guns, germs, and steel: the fates of human societies. W.W. Norton & Company

Fikri K, Juni Zhu T (2015) Competitive cities for jobs and growth. World Bank. https://documents1.worldbank.org/curated/en/323321467987876242/pdf/101718-REVISED-PUBLIC-CP1-Final-2.pdf

Glaeser E (2008) The triumph of the city. Penguin Random House

International Telecommunication Union (ITU) (n.d.-a) ITU's implementation of the U4SSC KPIs on smart sustainable cities. International Telecommunication Union. https://www.itu.int/en/ITU-T/ssc/Pages/KPIs-on-SSC.aspx

International Telecommunication Union (ITU) (n.d.-b) About international telecommunication union (ITU). International Telecommunication Union. https://www.itu.int/en/about/Pages/default.aspx

International Telecommunication Union (ITU) (n.d.-c) United for smart sustainable cities (U4SSC). International Telecommunication Union. https://u4ssc.itu.int/

International Telecommunication Union (ITU) (n.d.-d) Factsheets—united for smart sustainable cities (U4SSC). International Telecommunication Union. https://u4ssc.itu.int/factsheets/

International Telecommunication Union (ITU) (2016) Rep. ITU-T Y.4900/L.1600: Overview of key performance indicators in smart sustainable cities. International Telecommunication Union. https://www.itu.int/itu-t/recommendations/rec.aspx?rec=12627#:~:text=ITU%2DT%20Y.,-4900%2FL.&text=ITU%2DT%20Y.-,4900%2FL.,and%20Supplements%20that%20define%20KPIs

International Telecommunication Union (ITU) (2017) Rep. United 4 smart sustainable cities: collection methodology for key performance indicators for smart sustainable cities. International Telecommunication Union. https://www.itu.int/en/publications/Documents/tsb/2017-U4SSC-Collection-Methodology/files/downloads/17-00474_Collection-Methodology-for-Key-Performance-Indicators-for-Smart-Sustainable-Cities.pdf

International Telecommunication Union (ITU) (2021) Rep. Key performance indicators: a key element for cities wishing to achieve the sustainable development goals. International Telecommunication Union. https://www.itu.int/en/ITU-T/ssc/united/Documents/U4SSC%20Publications/KPIs-for-SSC-concept-note-General-June2020.pdf

Lozano-Gracia N, Soppelsa ME (2019) Pollution and city competitiveness. (World Bank Working papers) World Bank Group, February 2019. https://openknowledge.worldbank.org/server/api/core/bitstreams/cd85df27-afd1-5332-a471-5addee1d645d/content

UN Trade and Development (UNCTAD) (2022) Four key challenges facing least developed countries. UN Trade and Development. https://unctad.org/news/four-key-challenges-facing-least-developed-countries

UN-Habitat (2022) Rep. World cities report 2022: envisaging the future of cities. UN-Habitat. https://unhabitat.org/wcr/

United For Smart Sustainable Cities (U4SSC) (2020) City snapshot: Wels, Austria. United For Smart Sustainable Cities. November. https://www.itu.int/en/ITU-T/ssc/united/Documents/U4SSC%20Publications/City%20Snapshot/November%202020/U4SSC_Wels-Austria_City-Snapshot.pdf?csf=1&e=t5cZ7y

United Nations Economic Commission for Europe (UNECE) and International Telecommunication Union (ITU) (n.d.) Rep. United 4 smart sustainable cities. United Nations Economic Commission for Europe. https://unece.org/fileadmin/DAM/hlm/projects/SMART_CITIES/U4SSC-brochure.pdf

Wels Verwaltung (2021) Wels ist erste österreichische smart sustainable city. Wels Verwaltung. https://www.wels.gv.at/news/detail/wels-ist-erste-oesterreichische-smart-sustainable-city/

World Bank (2014) Urban development: sector results profile. World Bank. https://www.worldbank.org/en/results/2013/04/14/urban-development-results-profile

World Health Organization (WHO) (2020) Life expectancy and healthy life expectancy. Data by country. World Health Organization. https://apps.who.int/gho/data/node.main.688

Part II
Smart City Standardization: International Standards

ITU Standards for Smart Cities: Driving Digital Transformation

Hyoung Jun Kim

Abstract

The International Telecommunication Union (ITU), the United Nations agency for information and communication technologies (ICTs), has been actively working on developing standards on various aspects related to digital transformation in smart cities. Given the interdisciplinary nature of the concept of digital transformation, this paper elaborates on the various topic areas and key stakeholders that ITU engages with to formulate international standards in alignment with the Sustainable Development Goals (SDGs). The ITU-T Focus Group on Smart Sustainable Cities (FG-SSC) and ITU-T Study Group 20 have been instrumental in creating definitions and frameworks for smart cities, including the development of key performance indicators (KPIs) to assess progress towards sustainability and digital transformation. ITU-T standards cover various aspects of smart cities, including IoT, interoperability, scalability, security, and citizen-centric approaches. These standards facilitate knowledge sharing and transparency in the smart city marketplace. The document highlights the complexity of urban ecosystems and the need for digital transformation to improve efficiency and quality of life. It outlines three approaches to digital transformation in smart cities: technology-based, human-centric, and integrated. The ITU-T Study Group 20 has produced a range of standards across different series, addressing general principles, definitions, requirements, infrastructure, frameworks, services, management, identification, security, and evaluation. The KPIs developed by the ITU-T, particularly Recommendation ITU-T Y.4903, have been adopted by over 150 cities globally, including Daegu, Republic of Korea, to monitor progress towards SDGs. The paper

---

H. J. Kim (✉)
Electronics and Telecommunications Research Institute (ETRI), Daejeon, South Korea
e-mail: khj@etri.re.kr

also provides a strategic, process, and technical level guide for city leaders and planners to envision and implement digital transformation in alignment with ITU standards. It concludes by emphasizing the uniqueness of each city's journey towards smart city status and the importance of self-assessment using ITU-T KPIs rather than comparison to other cities.

Keywords

Smart cities · Digital transformation · ITU-T standards · Key performance indicators · United Nations

3.1 Introduction

Urban problems in cities are multifaceted and complex, reflecting the challenges of rapid urbanization, population growth, and the strain on infrastructure and resources. These issues are not new, but they have intensified with the pace of globalization and technological advancement. Among the most pressing urban problems are traffic congestion, air pollution, inadequate housing, insufficient public services, and the digital divide. Traffic congestion not only wastes time and fuel but also contributes to pollution and stress among commuters. Air pollution is a significant health hazard, leading to respiratory diseases and other health issues. Inadequate housing is a critical issue, with many urban residents living in slums or unaffordable housing, lacking basic amenities. The digital divide further exacerbates inequalities, as access to technology and the internet is not uniformly distributed across urban populations.

The concept of smart and sustainable cities has emerged as a promising solution to these urban problems. Smart cities leverage information and communication technologies (ICTs) to create more efficient, sustainable, and livable urban environments. They integrate various systems and services, such as transportation, energy, water, waste management, and public services, to optimize resource use and improve the quality of life for residents.

One of the key roles of smart cities is to address traffic congestion through intelligent transportation systems (ITS). These systems use sensors, cameras, and data analytics to manage traffic flow, reduce accidents, and provide real-time information to commuters. Smart parking solutions, for example, help drivers find available parking spaces, reducing the time spent circling in search of a spot. Similarly, smart public transportation systems can optimize routes and schedules based on demand, making public transit more efficient and attractive to users.

In terms of environmental sustainability, smart cities deploy technologies to monitor and manage energy and water consumption. Smart grids enable the efficient distribution of electricity, incorporating renewable energy sources and reducing waste. Smart water management systems detect leaks, monitor usage, and promote conservation, ensuring that water, a precious resource, is used wisely.

Smart cities also play a crucial role in addressing the issue of inadequate housing by promoting innovative urban planning and construction techniques. The use of modular and prefabricated construction can speed up building processes, reduce costs, and create more sustainable housing options. Additionally, smart cities can facilitate better public services by digitizing government processes, making them more accessible and transparent to citizens.

However, the future of smart cities also faces challenges, including cybersecurity threats, privacy concerns, and the need for inclusive policies to ensure that the benefits of smart technologies are accessible to all residents, especially vulnerable populations. Addressing these challenges will require collaboration among governments, the private sector, civil society, and citizens themselves.

The future of smart cities is bright, but it will require careful planning, ethical considerations, and a commitment to inclusivity to ensure that these innovations benefit everyone in the urban landscape.

The International Telecommunication Union's Telecommunication Standardization Bureau (TSB) commenced its work on standards in 2013, with the creation of the ITU-T Focus Group on Smart Sustainable Cities (FG-SSC). Following the completion of the activities of FG-SSC, the ITU-T Study Group 20 on "Internet of things (IoT) and smart cities and communities (SC&C)"was established in June 2015 to carry forth standardization work within this domain.

This Chapter traces ITU's body of work within the sphere of standardization and smart sustainable cities. After reading this chapter, the reader will gain insights into the importance of ITU-T standards, while delving into the key concepts of digital transformation in cities. The Chapter will also elaborate on key standards formulated by ITU-T on the topic of smart sustainable cities.

As city ecosystems are considered complex given the intricate mechanisms, ITU developed the Recommendation ITU-T Y.4903: Key performance indicators to assess smart sustainable cities transition in alignment with sustainable development goals. The process for the formulation of these KPIs involved an analysis of the various key areas in the urban ecosystem that can be impacted by the incorporation of technologies. As city ecosystems continue to evolve, these KPIs have been further revised, together with 19 United Nations agencies under the United for Smart Sustainable Cities (U4SSC) umbrella. U4SSC (United for Smart Sustainable Cities) is a global initiative that aims to promote the use of information and communication technologies (ICTs) to achieve sustainable urban development.

The rest of this chapter is organized as follows: Sect. 3.2 describes the organization and some background that affects its smart city standardization policy. Section 3.3 contains the information about the corresponding standards that have been developed by the ITU about Smart cities. Finally, Sect. 3.4 contains conclusions and future planning for smart city development.

3.2 Background

The International Telecommunication Union (ITU) is the United Nations specialized agency for information and communication technologies (ICTs). The ITU has three sectors:

ITU Radiocommunication Sector: This sector deals with the global management of the radio spectrum and satellite orbits.

ITU Telecommunication Development Sector: This sector's activities are dedicated to bridging the digital divide, especially in developing countries, with the aim of improving gender equality, sustainability etc.

ITU Telecommunication Standardization Sector: This sector develops international standards on key topics related to ICTs across domains. The main standards work within this sector is conducted through technical groups known as ITU-T Study Groups.

It has been estimated that by 2050, almost two thirds of the world's population will be living in urban areas, thereby increasing the population density of cities around the globe.[1] With the increase of the urban population, the development of infrastructure and services will become indispensable to meet the needs of the cities' inhabitants. Rapid urbanization has brought within its wake a plethora of challenges associated with urban sprawl, pollution, limited sources, land degradation. With the advent of the fourth industrial revolution, new and emerging technologies like the Internet of Things (IoT), Artificial Intelligence (AI), blockchain, metaverse are being embraced to tackle these urban challenges and establish people-oriented smart cities as they seek to improve quality of life (QoL) of its inhabitants while driving innovation.

However, despite the popularity of the concept of smart and sustainable cities, in the early 2010s, there was limited clarity on the use of the term.

The future of smart cities and technologies is promising, with ongoing advancements in artificial intelligence (AI), the Internet of Things (IoT), big data analytics, and blockchain. These technologies will enable even more sophisticated and responsive urban systems. AI, for instance, can analyze vast amounts of data to predict and prevent urban issues before they arise. The IoT will connect an ever-increasing number of devices, providing real-time data for better decision-making. Big data analytics will help city planners and policymakers understand complex urban dynamics and design more effective interventions. Blockchain technology can enhance security, transparency, and efficiency in various urban services, from land registries to energy trading. However the lack of clarity on the terminology remained a hurdle.

---

[1] https://www.un.org/en/desa/around-25-billion-more-people-will-be-living-cities-2050-projects-new-un-report.

Recognizing this, the International Telecommunication Union (ITU)[2] through its Focus Group on Smart Sustainable Cities (FG-SSC) developed the following definition, based on an analysis of over 100 related concepts, terms and definitions. Thus, the Focus-Group concluded to the following definition:

> A smart sustainable city is an innovative city that uses ICTs and other means to improve quality of life, efficiency of urban operation and services, and competitiveness, while ensuring that it meets the needs of present and future generations with respect to economic, social, environmental as well as cultural aspects.

This definition was later adopted by the United for Smart Sustainable Cities (U4SSC) initiative,[3] which is supported by 19 United Nations agencies and programmes.

To realize the global smart city dream, the process of digital transformation is pivotal as it focusses on the judicious integration digital technologies into various aspects of the urban ecosystem to enhance the way it operates, delivers value enhance customer experiences, enable innovation, and adapt.

As such in terms of driving digital transformation in cities, the process is dedicated to enhancing efficiency by automating operations and optimizing resource allocation, and exploring new avenues for innovation to deliver new services based on data-driven decision making and increased responsiveness to user needs across industries and empower citizens to actively participate in decision-making processes.

With the final aim of improving quality of life (QoL), initiating digital transformation in cities can have a cross-sectoral impact across all smart city verticals including health, transport, finance, retail, education to provide personalized services to inhabitants. In this context, digital technologies can facilitate the delivery of inhabitant-centric urban services. For example, telemedicine services can be provided to patients based in remote locations through existing channels such as mobile phones and smart wearables. In the post-COVID era, these services are invaluable given that the medical sector in several countries are still recovering from the burden of the pandemic. However, the increasing adoption of emerging technologies for healthcare also brings in personal data protection concerns which also need to be adequately addressed.

In this context, employing digital transformation for waste management sector can help monitor garbage levels, optimizing waste collection routes and schedules. On the sustainability front, digital transformation fosters energy efficiency and sustainability in smart cities. Smart grids, allow for real-time monitoring and optimization of energy usage and helps minimize environmental impact.

However, it's important to note that despite the benefits of digital transformation, its commencement in smart cities also raises concerns related to data privacy, cybersecurity,

---

[2] International Telecommunication Union (ITU) is the United Nations agency for information and communication technologies (ICTs) and is also an international standards developing organization (SDO).

[3] United for Smart Sustainable Cities: https://u4ssc.itu.int/.

and equitable access to technology. These challenges need to be addressed to ensure that the benefits of smart cities are accessible to all residents and that privacy and security concerns are managed.

While understanding where cities need to be in terms of optimally leveraging technologies, digital transformation needs to be undertaken in a planned and constructive way. In general terms, digital transformation is a process through which existing mechanisms within an organization or city can be enhanced by adopting digital technologies in response to changing needs of users. Digital transformation for the establishment of SSC will be focussed on boosting innovation, acquisition of digital skills, with the aim of generating new business models across sectors, with reduced operational costs and higher efficiency, in keeping with the needs of the inhabitants. Digital transformation (DT) in cities further facilitates the improvement of the existing services, along with the introduction of new ones to upgrade value chains and provide a new and better customer experience for inhabitants of a given city.

In cities, the implications and benefits of digital transformation, owing to the interconnected nature of city systems is difficult to predict. Nevertheless, the process of digital transformation, in cities will be oriented towards digital innovation, technology transfer, and sustainability.

In this context, it has been observed that urban stakeholders including (but not limited to) policymakers, utility providers, private sector entities and most importantly inhabitants have focused on three different approaches to digital transformation in smart cities:

1. Technology based approach: This approach involves emphasis on hardware and the adoption of technology-based infrastructure for the development (of a city). It has been observed that many developing countries continue to focus on this technology-based approach for infrastructural innovation as well as satisfying economic growth and achieving the SDGs. This technology or sector-based approach forms the stepping-stone for the integrated approach.
2. Human-centric approach: This approach involves investment in human and social capital for the establishment of smart cities. The supporters of this approach highlight the use of human/social capital as the basis of development of smart cities in terms of employing talent from sectors including science to engineering and design to research. This would include the involvement of the technology-centric entrepreneurs capable of developing innovative products and processes. This approach is essentially included in technology and integrated approach.
3. An integrated approach: This approach involves improving the quality of living of the inhabitants by integrating technological and social innovation. The integrated approach is based on the ability of cities to create conditions for continuous learning and innovation. This integrated approach to digital transformation smart cities is expected to be the most favored one in the times to come as it represents the view that merely enhancing the performance of individual sectors including transport energy, urban safety,

energy use, waste disposal with the help of digital technologies would not help build a smart city. Instead, smart cities are to be taken as a well-connected network of sectors in which digital applications may be used to achieve sustainable growth (economic, social, environmental).

3.3 Research Methodology: ITU-T Standards for Making Smart Sustainable Cities a Reality

In the digital age, the term "Smart Sustainable Cities" (SSC) has been the point of discussion to foster sustainable urbanization. However, a few decades ago, the world still grappled with trying to gain an understanding of this concept and its implementation. The lack of a proper and agreed definition for SSC made it difficult to analyse and apply the concept adequately. In this scenario, many cities self-designated themselves as smart cities, leaving major differences in the understanding and interpretation of the term among academia, and the public and private sector.

In 2014, the International Telecommunication Union (ITU), United Nations specialized agency for information and communication technologies (ICTs) overcame this problem by introducing the concept of "Smart Sustainable Cities" (SSC) which would include the major aspects of smart cities around the world as developed together with academics, city administrators, public sector entities and ICT companies.

The initial work on this was undertaken within its Focus Group on Smart Sustainable Cities (FG-SSC). Subsequently, the ITU-T Study Group 20 on "Internet of things (IoT) and smart cities and communities (SC&C)" was established in June 2015, to conduct studies related to Internet of things (IoT), machine-to-machine (M2M) communications, ubiquitous sensor networks and smart sustainable cities and its verticals. The work of ITU-T Study Group 20 also covers big data aspects of IoT, and smart cities, guidelines, methodologies and best practices related to cities, communities, rural areas and villages delivering services using emerging digital technologies, and interoperability of IoT and SC&C systems, services and applications.

In general, ITU-T standards related to smart cities (commonly known as ITU-T Recommendations) encompass relevant guidelines, protocols, best practices, architectural frameworks, use-cases for the deployment of smart city solutions. These standards can be leveraged by stakeholders for enhancing:

1. Interoperability: Smart cities consist of numerous interconnected systems, devices, and services from various vendors and providers. Standards facilitate interoperability, enabling these different components to seamlessly communicate, exchange data, and work together. Interoperability prevents the creation of silos, promotes integration, and allows for the efficient sharing and utilization of information, leading to improved efficiency and effectiveness of smart city solutions.

2. Scalability: Smart city standards provide a scalable framework that accommodates future growth and advancements. They ensure that smart city solutions can be easily expanded, upgraded, and integrated with emerging technologies without significant disruptions or costly modifications. Standards help cities avoid vendor lock-in and enable them to leverage new innovations as they arise.
3. Security: ITU-T Smart city standards incorporate security and privacy considerations to protect the sensitive data and infrastructure within a city. They establish best practices for data protection, encryption, access controls, and authentication mechanisms. By adhering to these standards, cities can enhance cybersecurity measures, mitigate risks, and build trust among citizens and stakeholders.
4. Citizen-Centric Approach: ITU-T smart city standards are oriented towards a citizen-centric approach, keeping the needs and preferences of residents at the forefront. These standards ensure that smart city solutions are designed and implemented with the aim of improving quality of life, promoting inclusivity, and addressing specific urban challenges.
5. Knowledge Sharing: ITU-T smart city standards are formulated based on close collaboration with 193 member states and over 900 private sector and academia industry players, and research institutions who put forth their experiences, and best practices. This collaboration enables cities to learn from each other, accelerate innovation, and avoid duplicating efforts.
6. Transparency in the marketplace: Given the wide range of stakeholders involved in the development of ITU smart city standards, their adoption provides a level of certainty and predictability for investors, vendors, and businesses operating in the smart city market. These standards create a stable and transparent environment that fosters trust and confidence, help expand the market and drive healthy competition and innovation.

The core part of ITU-T Study Group 20's work is focussed on the development of standards related to IoT technologies, architectures, protocols, and applications, to promote interoperability, security, and scalability in IoT deployments, frameworks, methodologies, and guidelines for the planning, deployment of smart city solutions across verticals including education, data management, mobility, etc. To avoid duplication of work, Study Group 20 coordinates with other ITU groups, relevant organizations, and industry stakeholders to ensure harmonization and coordination of standards development efforts.

The types of standards it develops include those listed in Table 3.1.

To further gain an in depth understanding of driving digital transformation in cities, ITU-T Study Group 20 is developing a definition on "Digital transformation for people-centred smart cities and communities" by analyzing different definitions related to the topic.

Table 3.1 List of standards developed by ITU-T study group 20

Series	Examples
Y.4000–Y.4049: General	• Y.4000: Overview of the internet of things • Y.4001: Machine socialization: overview and reference model • Y.4002: Machine socialization: relation management models and descriptions • Y.4003: Overview of smart manufacturing in the context of the industrial internet of things
Y.4050–Y.4099: Definitions and terminologies	• Y.4050: Terms and definitions for the internet of things • Y.4051: Vocabulary for smart cities and communities
Y.4100–Y.4249: Requirements and use cases	• Y.4204: Accessibility requirements for the internet of things applications and services • Y.4205: Requirements and reference model of IoT-related crowdsourced systems • Y.4206: Requirements and capabilities of user-centric workspace service • Y.4207: Requirements and capability framework of smart environmental monitoring
Y.4250–Y.4399: Infrastructure, connectivity and networks	• Y.4251: Capabilities of ubiquitous sensor networks for supporting the requirements of smart metering services • Y.4252: Energy saving using smart objects in home networks
Y.4400–Y.4549: Frameworks, architectures and protocols	• Y.4401: Functional framework and capabilities of the internet of things • Y.4402: Requirements and functional architecture for the open ubiquitous sensor network service platform • Y.4403: Functional requirements and architecture of the next generation network for support of ubiquitous sensor network applications and services • Y.4404: Framework of object-to-object communication for ubiquitous networking in next generation networks

(continued)

Table 3.1 (continued)

Series	Examples
Y.4550–Y.4699: Services, applications, computation and data processing	• Y.4551: Service description and requirements for multimedia information access triggered by tag-based identification • Y.4552: Application support models of the internet of things • Y.4553: Requirements of smartphone as sink node for IoT applications and services • Y.4554: Requirements for network-based location information conversion for location-based applications and services
Y.4700–Y.4799: Management, control and performance	• Y.4700: Deployment guidelines for ubiquitous sensor network applications and services for mitigating climate change • Y.4701: SNMP-based sensor network management framework • Y.4702: Common requirements and capabilities of device management in the internet of things
Y.4800–Y.4899: Identification and security	• Y.4809: Unified internet of things identifiers for intelligent transport systems • Y.4810: Requirements for data security of heterogeneous internet of things devices • Y.4811: Reference framework of converged service for identification and authentication for IoT devices in a decentralized environment
Y.4900–Y.4999: Evaluation and assessment	• Y.4901: Key performance indicators related to the use of information and communication technology in smart sustainable cities • Y.4902: Key performance indicators related to the sustainability impacts of information and communication technology in smart sustainable cities • Y.4903: Key performance indicators for smart sustainable cities to assess the achievement of sustainable development goals

3.3.1 ITU Smart City KPIs

Cities are also expected to play a pivotal role in attaining the Sustainable Development Goals (SDGs). It is important to understand that urban and rural development are not mutually exclusive as cities provide inputs as well as output markets for rural areas. However, many cities find themselves saddled with low economic productivity, unemployment, expanding slum areas, environmental degradation etc. For countries to meet the SDGs, it is essential to develop long-term strategies for the health, education, mobility, urban planning, tourism—among others. This is where digital technologies would serve as an indispensable resource. Aligning smart city goals with the SDGs, are an ideal means to achieve both these sets of targets simultaneously.

To underscore the potential role that SSCs could play in achieving the SDGs, it is importance to comprehend that just as the SDGs cannot be attained overnight, the cities cannot transform into SSC within a day. In this context the smart and sustainable transitions need to be viewed as a long-term journey. No city in the world is alike. Therefore, cities to be able to measure their progress with reference to their smart city goals. In this context, there needs to be an effective tool to monitor advancements and for city's conduct self-assessments in relation to SDG targets.

Realizing this the ITU developed the Recommendation ITU-T Y.4903: Key performance indicators for smart sustainable cities to assess the achievement of sustainable development goals. These KPIs were further revised under the framework of U4SSC, together with partner UN agencies and based on feedback from cities. As of 2022, over 150 cities worldwide have already implemented these KPIs. In 2022, the city of Daegu, Republic of Korea also adopted these KPIs to assess the city's progress towards the 169 targets embedded in the SDGs. This journey was documented in a case study on Daegu's strategic approach to urban development, focusing on industrial growth and citizen happiness. The city has implemented various smart initiatives, including the Suseong Alpha City testbed for autonomous driving and smart streetlights, and plans to integrate and utilize city-wide data through a Smart City Data Hub. Daegu's Living Lab System engages citizens in identifying and solving urban issues, contributing to the city's 2021–2025 Smart City Plan.

The case study serves as a valuable reference for cities seeking to develop their smart city strategies, providing a comprehensive overview of Daegu's approach, the implementation of the U4SSC KPIs, and the city's vision for a smart sustainable future (Fig. 3.1).

The KPIs have been divided into three core dimensions: economy, environment and society and culture. Some examples of U4SSC Key Performance Indicators include (but not limited to):

1. Energy: The related KPIs focus on measuring electricity consumption, renewable energy generation, energy efficiency, and greenhouse gas emissions. Examples of energy-related KPIs include energy intensity, share of renewable energy.
2. Water and Sanitation: Water-related KPIs assess the management and conservation of water resources. This includes indicators such as water consumption per capita, water loss in the distribution network, and the proportion of wastewater treated before discharge.

Fig. 3.1 Overview of Daegu's implementation of the U4SSC KPIs (ITU 2021)

3. Waste: These KPIs for waste management measure the effectiveness of waste collection, recycling, and disposal practices. Examples include waste generation per capita, recycling rate, and the amount of waste disposed of in environmentally friendly ways.
4. ICT Infrastructure: KPIs related to ICT infrastructure assess the availability, quality, and accessibility of ICT services in cities. This includes indicators such as broadband coverage, internet access speed, and the percentage of the population using digital services.
5. Social Inclusion: Social inclusion KPIs focus on measuring the inclusiveness and accessibility of services for all segments of the population. This includes indicators such as access to education, healthcare services, affordable housing, and digital inclusion.

With the help of these indicators cities worldwide have been able to assess their progress, benchmark best practices, while tracking the impact of their smart sustainable city initiatives.

3.3.2 Strategic Level

It is not easy for a city to commence its smart city journey towards digital transformation. Given the complex nature of urban ecosystems, cities would need additional guidance to better comprehend their strengths and weakness. Accordingly, the Guide for Smart and Sustainable City leaders: Envisioning Sustainable Digital Transformation (based on Y.Sup32: ITU-T Y.4000 series—Smart sustainable cities—A guide for city leaders) was developed to highlight a general eight-step cycle for digital transformation in aspiring smart cities and underscoring relevant international standards for implementation.

3.3.3 Process Level

The Collection Methodology for Key Performance Indicators for Smart Sustainable Cities (developed under U4SSC) provides cities with a methodology on how to collect relevant data or information from different sources to assess their goals with reference to their smart city goals.

This allows for collecting relevant data in reference to the KPIs which are designed to help cities assess their progress towards the Sustainable Development Goals (SDGs), becoming smarter and more sustainable. They cover three dimensions—Economy, Environment, and Society and Culture—and are subdivided into core and advanced indicators. Core indicators are basic measures achievable by all cities, while advanced indicators provide deeper insights but may be beyond the capabilities of some cities.

The KPIs are structured to enable cities to measure performance over time, compare with other cities, and share best practices. They are categorized under specific areas such as ICT Infrastructure, Water and Sanitation, Energy Supply, Transport, Public Sector, Innovation, Employment, and more. Each KPI is chosen through a rigorous process involving international experts and UN agencies, ensuring relevance to the SDGs and comparability.

The methodology provides detailed instructions for collecting data, including rationale for choosing indicators, interpretation, benchmarking, data sources, and calculation methods. It emphasizes the importance of consistent and standardized data collection to support meaningful comparisons and the development of the U4SSC Smart Sustainable City Index.

3.3.4 Technical

The Master Plan—Enabling Digital Transformation in Smart Cities (based on Y.Sup33: ITU-T Y.4000 series—Smart sustainable cities—Master plan) provide an overarching roadmap of how digital technologies can be leveraged for smart sustainable cities in alignment with ITU standards for the formulation of tailored digital transformation strategies.

3.4 Conclusions

ITU-T standards (also referred to as ITU-T Recommendations) are developed within ITU-T Technical Groups known as Study Groups. There are currently 11 active study groups, each with its own specific scope. The standards are developed in a spirit of consensus with help of 193 member states and over 900 private sector and academia members who contribute to the standards development work.

No two cities are alike given their geographic location, social fabric and governance. Every city will start at different stage of digital transformation or smart city transition. This means that it is not unfair to pit cities against each other with reference to their smart city initiatives. In this scenario, how are cities to monitor progress or advancement. Without competing with each other, cities can monitor their own annual progress in alignment with their own individual smart city targets with the help of U4SSC KPIs.

ITU has played a pivotal role in driving the digital transformation of smart cities through the development of international standards that are aligned with the Sustainable Development Goals (SDGs). The ITU-T Study Group 20 has been at the forefront of this effort, creating a comprehensive set of standards that address the complex and multifaceted nature of smart city ecosystems. These standards not only ensure interoperability, scalability, and security but also prioritize a citizen-centric approach, thereby enhancing the quality of life for urban inhabitants.

The KPIs developed by ITU-T, as outlined in the Recommendation ITU-T Y.4903, have become an invaluable tool for cities around the world to measure their progress towards becoming smart sustainable cities. By revising these KPIs in collaboration with UN agencies and other stakeholders, the ITU has provided a robust framework that allows cities to benchmark their achievements against the SDGs, fostering a global movement towards sustainable urban development.

The strategic, process, and technical level guidance provided by the ITU, including the Guide for Smart and Sustainable City leaders and the Master Plan for enabling Digital Transformation in Smart Cities, offer a roadmap for cities to navigate the challenges of digital transformation. This guidance is essential for cities to leverage digital technologies effectively, optimize resource allocation, and drive innovation in a manner that is inclusive and sustainable.

As the world continues to urbanize, the importance of the ITU's work in standardizing smart city technologies cannot be overstated. By facilitating knowledge sharing, promoting transparency in the marketplace, and fostering trust among stakeholders, the ITU-T standards contribute to a stable environment that encourages healthy competition and innovation.

ITU's commitment to creating a consensus-driven framework for smart city development is a testament to its role as a global leader in the field of information and communication technologies. The organization's collaborative approach, which brings together member states, industry players, and academia, ensures that the standards developed are comprehensive, relevant, and effective in guiding the transformation of cities into smart, sustainable, and inclusive urban environments.

Acknowledgements We would like to thank the International Telecommunication Union (ITU), especially experts from ITU-T Study Group 20 on IoT, Smart Cities and Communities. I would also like to extend my deepest appreciation to the SG20 Management Team, Rapporteurs and Editors for their hard work in driving the standardization work forward. I would also like to thank Seizo Onoe, Director of the Telecommunication Standardization Bureau, ITU.

References

ITU (2021) City Snapshot—Daegu, Korea (Republic of). Retrieved from https://www.itu.int/dms_pub/itu-t/opb/tut/T-TUT-SMARTCITY-2021-37-PDF-E.pdf

Dr. Hyoung Jun Kim began his career in 1988 at the Electronics and Telecommunications Research Institute (ETRI) where he currently holds the position of Research Fellow. Previously, he served as Senior Vice-President and led the Intelligent Convergence Research Lab at ETRI. Over the course of 38 years, Dr. Kim has gained research experience in a range of ETRI divisions, such as the Info-Communications Technology Division, IT Strategy Research Division, Information and Telecommunications Technology Division, and Protocol Engineering Centre. Presently, he also serves as the Chairman of both ITU-T SG20 and the APT Standardization Program (ASTAP) within the Asia–Pacific Telecommunity, and Vice Chairman of Network of Women in ITU-T. In the past, he has held roles as Vice-chair of SG20 and SG13, Rapporteur of Q25/16 since 2004, Vice-Chairman of the Focus Group on M2M Service Layers, Chairman of WP2 of the same Focus Group in ITU-T SG11, and Vice-Chairman of the Focus Group on Future Networks in ITU-T SG13. Dr. Kim's research and academic achievements are extensive, with over 450 standard proposals, more than 150 academic journal and conference papers, over 100 patents, and 20 official technology transfers to domestic companies under his belt. He has received three National President's citations in 2003, 2009 and 2023, in addition to numerous Certificates of Appreciation from international standard-related organizations, including ITU-T. Notably, Dr. Kim earned his second and third National President's Award at the "World Standards Day 2009 in Korea and 2023 in Korea" in recognition of his dedication to international standardization.

Part III
Smart City Standardization: European Efforts

4

ETSI Standards for Smart Cities: Standards for Interoperable, Sustainable and Accessible Citizen-Centric Services

Laure Pourcin

Abstract

Smart cities are about creating a better place to live for people. They use Information and Communications Technologies (ICT) to optimize and adapt their resources and services to the evolving needs of their citizens. This chapter gives an overview of the challenges facing their development and how global standards can provide the foundation for the creation of interoperable, sustainable and accessible people-centric smart cities. As a provider of world-class standards for ICT, ETSI has been involved in developing standards for smart cities for over 10 years now, involving a variety of stakeholders of the ecosystem into the development of these standards.

Keywords

Interoperability · Citizen-centric · Sustainability · Accessibility · Standardisation

4.1 Introduction

Above all, smart cities should be about creating a better place to live for people. One might define smart cities as cities that use Information and Communications Technologies (ICT) to optimize and adapt their resources and services to the evolving needs of the people. The urban context concentrates in a single location a great number of people involved in diverse activities. Cities are generally included and interlinked in wider communities, including urban, suburban and rural areas. The functioning of the city and community

---

L. Pourcin (✉)
European Telecommunications Standards Institute (ETSI), Valbonne, France
e-mail: Laure.Pourcin@etsi.org

© The Author(s), under exclusive license to Springer Nature Switzerland AG 2025
L. Anthopoulos (ed.), *Smart City Standardization*, Synthesis Lectures on Computer Science, https://doi.org/10.1007/978-3-031-95959-2_4

services involves many different stakeholders in a broad variety of use cases, such as the management of traffic flows, lighting, public transport, environmental conditions, health and emergency services, events, security, etc.

The purpose of this chapter is to give an overview of the challenges facing the development of smart cities and how global standards can provide the foundation for the creation of interoperable, sustainable and accessible people-centric smart cities and communities. Section 4.2 provides background information for the development of smart and sustainable cities and communities, highlights the benefits derived from the use of global standards in smart cities and communities, and identifies the main drivers behind the development of smart city standards. Section 4.3 presents a few examples of use cases to illustrate the potential for innovation in developing citizen-centric services and provides an overview of the most significant standards published by ETSI in this domain. It also gives a glimpse of anticipated future standards developments for smart cities and communities in ETSI.

As a provider of world-class standards for Information and Communication Technologies (ICT), ETSI has been involved in developing standards for smart cities for over 10 years now. While nearly all standards developed by ETSI are implemented in cities, such as standards defining communication technologies (4G and 5G mobile, DECT, etc.), this article will focus on the standards which are of specific interest to the city's public authorities and private service providers when designing and implementing their smart city strategy and services.

4.2 Background

4.2.1 Smart City Challenges in the Twenty-First Century

Today more than half of the world's population lives in urban areas, a figure that is projected to grow to nearly 70% by 2050 according to the United Nations (UN). Combined with the impact of climate change on the environment, this trend places new demands on key city services and infrastructure such as transport, energy, health care, water and waste management.

The **United Nations Sustainable Development Goal 11** (UN 2015) is to make cities and human settlements inclusive, safe, resilient and sustainable. For all of us to survive and prosper, we need new and intelligent urban planning that creates safe, affordable and resilient cities with green and culturally inspiring living conditions. This goal encompasses issues such as housing, transportation, urbanization, heritage protection, natural disasters, environmental impact, green spaces, development planning, resource efficiency, and sustainable and resilient building.

To achieve these goals, a smart city can leverage Information and Communications Technologies (ICT) by integrating the requirements of its urban community, in terms of energy and other utilities (production, distribution and use), environmental protection, mobility and transport, services for citizens (healthcare, education, emergency services, etc.) and with proper regard for security, both of individuals and their personal data, and use it as a driver for economic and social improvements.

4.2.2 The Benefits of Global Standards

Having to cope with existing heterogeneous and siloed legacy systems and recognizing the need to ensure commonalities between the approaches taken by the different application areas of the smart city ecosystem, standards were developed to enable the cities to derive the best horizontal advantage from an overall approach to achieve interoperability of their services. ICT standards also define quality features of services and products, including security, safety, resilience, and sustainability of digital solutions to be deployed in smart cities and communities.

In this context, global standards published by ETSI and International Standardisation Organizations such as the International Standards Organization (ISO) and the International Electrotechnical Commission (IEC), provide the technical foundations to reach the following objectives for smart cities and communities:

- engage more actively with citizens and enhance the well-being of the citizens,
- enable the interoperability of services and enhance the city's performance,
- improve the sustainability and reach CO_2 emission reduction goals,
- provide accessibility and inclusivity for all,
- reduce operational costs and the city resource consumption,
- generate new business opportunities and increase the attractiveness of the city,
- ensure security, protection of privacy and safety.

Besides improving the well-being of citizens and visitors of the city, this harmonization leads to:

- cost reductions by avoiding vendor lock-in in proprietary solutions,
- better efficiency in the use of scarce resources, such as for instance energy and water,
- increased resilience, and
- innovation opportunities for local players including small and medium enterprises.

4.2.3 Drivers Behind the Smart City Standardisation Effort

The **United for Smart Sustainable Cities (U4SSC)** is a global UN initiative coordinated by the International Telecommunications Union (ITU), the United Nations Economic

Commission for Europe (UNECE) and UN-Habitat. U4SSC provides an international platform for information exchange and partnership building to guide cities and communities in achieving the UN Sustainable Development Goals.

Networks and associations of cities took a major role in triggering the standardisation effort at national, European and global level. Among the most active are the international network of **Open and Agile Smart Cities (OASC)** and **EUROCITIES** a community of over 200 cities in 38 countries and 130 million inhabitants. Their effort was complemented by national associations such as the NGO eG4U (e.Green for users) in France or the FEMP (Federación Española de Municipios y Provincias) in support of the development of open data publication in Spain, to cite only two examples.

The sharing of knowledge across borders enabled cities to familiarise themselves with the development and implementation of standards and to take an active role in the relevant Standardisation bodies. While facilitating the adoption of standards by all players in the city ecosystem, this involvement also contributed to the dissemination of best practices and innovative use cases among city public authorities and private service providers.

In Europe, this standardisation effort was reinforced and supported by several important projects funded by the European Union, including ESPRESSO looking for an open and interoperable platform for smart cities, and SYNCHRONICITY aiming to establish a reference architecture for IoT-enabled city applications. Today, sustainable smart cities and communities continue to be an integral part of the European Union rolling plan for ICT standardisation and standards development work in this domain continues to receive full support at EU level.

4.3 Standards for Smart Cities

4.3.1 Standardisation Ecosystem for Smart Cities

As indicated above, the main players in the standardisation ecosystem for smart cities are (i) the Standards Development Organisations (SDO), (ii) stakeholders' associations, and (iii) policy initiatives to support research, pilot or implementation programs. Below is a non-exhaustive list of the main players that have contributed to the development of standards for Smart Cities.

The **International Organization for Standardization (ISO)** is a worldwide federation of national Standardisation bodies dedicated to developing international standards through technical committees specialized in a particular field. Together with the **International Electrotechnical Commission (IEC)**, they publish international standards for worldwide use. Among the first published standards for Smart Cities, ISO 37120:2018 (ISO 2018) establishes definitions and methodologies for indicators for city services and quality of life, and ISO/IEC 30182:2017 (ISO 2017) defines the Smart City Concept Model (SCCM) that can be used to describe data from a smart city to facilitate interoperability.

Created in 1988 as the European Telecommunications Standards Institute, **ETSI** provides members with an open and inclusive environment to support the timely development, ratification and testing of globally applicable standards for ICT-enabled systems, applications and services across all sectors of industry and society. ETSI is a not-for-profit body with more than 900 member organizations worldwide, drawn from 64 countries and five continents. Members comprise a diversified pool of large and small private companies, research entities, academia, government and public organizations.

ETSI is one of only three bodies officially recognized by the European Union as a **European Standards Organization (ESO)**, together with the European Committee for Standardisation (CEN), and the European Committee for Electrotechnical Standardisation (CENELEC).

The need for developing global standards in support of the development of Smart Cities was identified very early on in ETSI. They were developed in various ETSI Technical bodies to cover different aspects, specifically in ETSI Technical Committee SmartM2M (TC SmartM2M), Industry Specification Group on Context Information Management (ISG CIM), Technical Committee on Environmental Engineering (TC EE), Technical Committee on Human Factors (TC HF), Technical Committee on Access, Terminals, Transmission and Multiplexing (TC ATTM), and Industry Specification Group on Operational energy Efficiency for Users (ISG OEU). These standards are presented in Sects. 3.3, 3.4, 3.5 and 3.6 below.

The 3rd Generation Partnership Project (3GPP) (3GPP 1998) unites seven telecommunications standard development organizations (the Association of Radio Industries and Businesses in Japan ARIB, the Alliance for Telecommunications Industry Solutions in the USA ATIS, the China Communications Standards Association CCSA, ETSI in Europe, the Telecommunications Standards Development Society in India TSDSI, the Telecommunications Technology Association in Korea TTA, and the Telecommunication Technology Committee in Japan TTC), known as "Organizational Partners" providing their members with a stable environment to produce the Reports and Specifications that define 3GPP technologies. 3GPP specifications cover cellular telecommunications technologies, including radio access, core network and service capabilities, which provide a complete system description for mobile telecommunications. The 3GPP specifications also provide hooks for non-radio access to the core network, and for interworking with non-3GPP networks.

oneM2M (2012) partnership was launched in 2012 as a global initiative between eight of the world's preeminent standards development organizations: ARIB, ATIS, CCSA, ETSI, TIA, TSDSI, TTA, TTC to develop specifications that ensure the most efficient deployment of Machine-to-Machine (M2M) communications systems and the Internet of Things (IoT). oneM2M Standardisation work continues today as a joint effort of TIA, TSDSI, TTA, CCSA and ETSI, with the active involvement of organizations from M2M-related business domains such as: telematics and intelligent transportation, healthcare, utilities, industrial automation, smart homes, public safety and services, and smart cities.

The **AIOTI** Association (2015) is The **Alliance for Internet of Things and Edge Computing Innovation**, founded by the European Commission in 2015. Currently consisting of ten working groups, AIOTI has a Working Group on Standardisation. AIOTI past and current work contributed IoT solutions to enhance city performance, safety and well-being, reduce cost, and increase citizen engagement and active participation. AIOTI has an interest in city-related sectors like mobility, energy, healthcare, manufacturing and agriculture, among others.

The **ESPRESSO** H2020 project (European Commission 2018) analysed different sectorial systems to develop a conceptual Smart City Information Framework based on open standards, including Minimum Interoperability Mechanisms. The project analysed the core vocabularies for smart cities, which enable cities and stakeholders to share data in an interoperability way in different smart city sectors.

The **SYNCHRONICITY** project (2018) aimed to establish a reference architecture for IoT-enabled city applications with identified interoperability points, interfaces and data models for different verticals, through the implementation of pilot projects at scale.

As stated above, it is nearly impossible to provide an exhaustive list of all relevant stakeholder's associations and public initiatives. The list above is limited to those which directly contributed to Standardisation work in ETSI. Hopefully, the reader will find more comprehensive information in the other chapters of this book.

4.3.2 Smart City Use Cases Examples

Use cases are used in standardisation to identify the requirements that will be met in the technical specifications. As mentioned in the introduction, there are almost infinite possibilities of use cases, combining the various services, resources and infrastructures of the city to adapt to the needs and improve the life of the people. The key to a successful use case is to put the citizens at the heart of the development of new services.

The examples below are only used as illustration of the multiple possibilities, they are taken from ETSI TR 103 506 and ETSI GR CIM 002 and derived from previous publications and pilots. We invite our readers to refer to ETSI standards and project publications for more use cases (Fig. 4.1).

(a) eHealth and assisted living at home (Source: ETSI TR 103 506 (ETSI 2018c))

Taking care of an aging population is one of the major challenges of the future of healthcare. An important step is the need to move from institutional care to assisted-living at home, in particular for elderly people living alone and for people with long-term needs and chronical illness (such as people with hypertension, dementia, or obesity). Electronic medical care services enable these people to obtain a better quality and independent life.

By making use of different data coming from monitoring equipment, wearables and building sensors, the goal is to provide assisted living to elderly people and people with

Fig. 4.1 ETSI illustration of citizen-centric smart city use cases (ETSI 2018c)

long-term needs. Actors in this use case include not only the monitored people and their close circle (relatives, neighbours), but also health service providers (ambulances, doctors, pathologists, nutritionists), assistance services (call centres), and the same municipality. The data would be collected from sensors from the building and smart home domains, for example motion sensors, occupancy sensors, or pressure mats, among others.

The collected data would be used by added-value services to detect abnormal conditions in elderly people's daily life environment and to trigger alarms to the municipal 24-h call centre where specialists can provide assistance. In addition, the use case is complemented with smart allocation of parking space when needed for municipality health services in case their actuation is needed in a particular location, for example when a user presses the panic button.

(b) Street lighting, air quality monitoring and mobility (Source: ETSI TR 103 506 (ETSI 2018a)**)**

Street lighting is a key city infrastructure for citizens as it is directly associated by them to the sense of security at night. However, this idea of security in addition to the sense of wealth and modernity brought by lighting systems, is increasing the light pollution levels mostly in highly populated cities. On the one hand, both the awareness about the light pollution consequences over citizens' health and the need for reducing costs in energy are pushing local administrations to find solutions for a more efficient street lighting system.

On the other hand, city administrations should handle the well-known problem of air pollution suffered also by big cities due to the high concentration of vehicles emitting CO_2.

Thirdly, there is also a need for reducing pedestrian accidents due to maintenance works and issues on the roads. Taking all this into account, this use case aims to:

- balance the energy spent on street lighting while keeping roads safe for pedestrians,
- re-route traffic to reduce high concentrations of air pollution in specific points,

- provide alternative routes for pedestrians to avoid areas with air pollution and potential hazards due to road works or issues.

For this, local administrations need to provide and deploy the infrastructure to monitor traffic and manage the street lighting configuration. For example, there is a need for sensors to detect presence to reduce the intensity of lighting when there is no traffic or pedestrians and increase it otherwise. Citizens could register road incidences by means of mobile or web applications as well as city reporting services. Such information could be used for suggesting safer routes to pedestrians. Final users include drivers, who could be re-directed to different routes according to the air pollution concentration in specific points, or pedestrians and cyclers who could be re-directed according to the safety and pollution conditions of the roads.

(c) **Crowd monitoring and emergency response (Source: ETSI GR CIM 002 ((ETSI 2023a, b)))**

Crowd/Queue detection is important in modern society to manage public places and vacation resorts for different situations, especially in case of emergency. The results of crowd/queue detection need to be combined with data from other sources to support knowledge discovery, decision making and smart recommendations. In an overcrowded place, when a crisis or an emergency occurs, the population needs to be evacuated while the injured people need to be transported to the hospital.

This case can for example apply to a vacation resort, major event facility or entertainment park. In the large vacation resort, many facilities exist for entertainment. To improve the vacationers' experience, queue detection is applied in front of entertainment facilities, while crowd detection can be applied in internal/external areas to detect crowd level. The data from queue/crowd detection is used by resort management system to provide real-time information, make smart recommendations for vacationers, propose special offers and respond to emergencies. The local authorities also use the data on a city scale to study the behaviour of vacationers in the resort and tourists in the city.

Devices for crowd detection, installed in different areas of the place, provide crowd information and can be accessed by other services, while the place management system provides information on available resources and real-time information relating to the location. The police department obtains a range of data, including the status of the emergency situation, traffic and road congestion in the vicinity, and information about the location and number of people. This is required to evacuate people from the crisis scene. Medical responders obtain information about available health care resources (hospitals, ambulances, doctors), and about traffic conditions. This is required to organize first aid and to transport the injured to available hospitals. IoT devices (e.g. cameras, sensors) and smartphones allow the collection of location and status data about people. Other systems can provide information on the incident, disposition of emergency response teams, equipment and facilities.

These are only three examples of use cases. Some were illustrated in a humoristic way in the ETSI video on "Smart Cities made simple" (ETSI 2018c). Many more are yet to be invented, to adapt to the changing context and needs of citizens. The application scope of Smart Cities is very broad covering parking, health, noise and traffic congestion maps, lighting, waste, transport systems, green mobility, residential energy efficiency and much more. Provided the right technical enablers are put in place, innovative use cases may arise and develop, improving the lives of millions of people living in urban areas.

4.3.3 Approach to Standardisation for Smart Cities

The creation of sustainable citizen-centric smart cities can only be achieved with a holistic approach in a long-term process, supported by global standards that enable fully interoperable solutions to be deployed and replicated at scale.

To reap the benefits of global standards for smart cities highlighted in Sect. 2.2 above, ETSI Technical Committees are developing specifications for sustainable smart cities and communities in three main technical domains:

- Standards for interoperable machine-to-machine (M2M) communications, data and applications, presented in Sect. 3.4 below
- Standards for green and sustainable smart cities and communities, presented in Sect. 3.5 below
- Standards for user-centred, safe and accessible services and facilities, presented in Sect. 3.6 below

Section 3.7 introduces the anticipated directions for future studies and standards development in ETSI.

Standards are of different types, some are normative, others are informative. The normative ones describe precisely a technology/data/architecture to be implemented, including the criteria to verify the compliance of a given solution's implementation with such standard. Other standards are informative, they are used to describe possible use cases, to provide guidance on how to implement the normative standards in practice, or for any other purpose. A set of standards has also been developed for the specific purpose of monitoring and measuring Key Performance Indicators (KPIs) of sustainable and smart cities and communities.

The use of standards is voluntary, ETSI published standards are available from ETSI's website free of charge to any entity or organization interested in implementing them. ETSI strongly recommends to city authorities and service providers alike, to reference and implement the relevant global standards in their purchasing and delivery processes. This good practice will safeguard their investment, ensure the state-of-the-art performance and interoperability of their digital solutions.

4.3.4 Standards for Interoperable Smart City Services

Open and standardized interfaces for products of different brands to interoperate are necessary to avoid vendor-lock in. Interoperability offers the business benefit to unlock new added value services for consumers from data integration, while manufacturers and other commercial parties can still maintain their competitive advantage in offering their solutions (not everything needs to become open and interoperable).

Machine-to-Machine (M2M) communications form the foundation layer for a fast-evolving world of smart devices, appliances, homes, cities and communities. Nowadays, smart cities provide important interoperability use cases for Internet of Things (IoT) applications, since they are by default requiring cross-domain interworking.

ETSI TC SmartM2M provides (together with oneM2M) a comprehensive standards-based solution including, among others, **IoT Semantic Interoperability**. **The Smart Application REFerence ontology (SAREF)** enables smart devices to exchange information using this common conceptualization of the reality. TC SmartM2M is in the process of publishing ETSI EN 303 760 (ETSI 2024a, b, c), a guide on how to develop, apply and evolve smart applications ontologies using SAREF. This standard provides the following explanation.

In the past, interoperability used to be addressed at the technical communication level. In recent years, the interoperability challenge has been raised to the information level, where the common concepts for all existing data models/protocols can be incorporated in an ontology (i.e., a common vocabulary). This captures the meaning of a concept (i.e., semantics) rather than the specific data format in which the concept is encoded for data exchange at the underlying communication layer.

The SAREF ontology developed and maintained by ETSI since 2015 provides a mature, sustainable and standardized framework of ontologies for IoT that enables different parties to interoperate with each other at the semantic level. The SAREF ontology keeps evolving and expanding to address the requirements of new use cases. Referring to the requirements identified from the use cases in ETSI TR 103 506 (ETSI 2018a), TC SmartM2M has specified an extension of the SAREF semantic model for the smart cities domain called SAREF4CITY in ETSI TS 103 410-4 (ETSI 2019a, b). This deliverable only covers the ontology categories that are specific to smart cities, such as administrative areas, city object, public service and topology.

As already mentioned, sustainable smart cities and communities provide multi-domain services and therefore the whole SAREF core and extension ontology standards (ETSI TS 103 410 multiple parts, defining ontology extensions to the SAREF model in different application domains), are relevant to smart cities:

- Energy domain
- Environment domain
- Building domain

- Smart City domain
- Industry and manufacturing domain
- Smart agriculture and food chain domain
- Automotive domain
- eHealth/Ageing-well domain
- Wearable domain
- Water domain
- Lifts domain
- Smart grid domain

The official ETSI Portal for SAREF gives user communities direct access to SAREF ontologies and related work items, allowing stakeholders to share their specific requirements and give their direct feedback on their use of SAREF ontologies.

To create smart citizen-centric services, it is also essential to accurately record data together with its context information (space, time, relations) and to transfer them without misinterpretation to other systems, in other words ensure **data interoperability**.

ETSI Industry Specification Group on cross-sector Context Information Management (ISG CIM) develops technical specifications and reports to enable multiple organisations to develop interoperable software implementations of a cross-cutting Context Information Management (CIM) layer, for smart cities applications and beyond. For this purpose, ETSI ISG CIM have developed the **NGSI-LD information model and corresponding APIs.**

To ensure data interoperability, it is necessary to establish data governance, management, and identification systems. Before establishing a data interoperability system, a policy or legal foundation for data interoperability should be established through consultation with policy-related parties through data governance. In addition, it is important to match the data format to share data between different platforms or domains, as well as to manage the lifecycle of the data by applying an integrated management system. In particular, using a common interface system in data lifecycle is essential to provide data interoperability between different platforms.

The NGSI-LD Information model can express relationships between data and can use the property graph model to express entity, property, and relationship of data. In addition, NGSI-LD defines standardized RESTful APIs for creating, retrieving, updating, and deleting context data, theses APIs facilitate the interaction between the system and applications in a consistent manner. Lastly, NGSI-LD employs ontologies and semantic annotations to ensure a shared understanding of data. This helps in aligning data from different sources and systems. Therefore, in order to secure data interoperability in smart cities, NGSI-LD can connect data between different platforms and domains, so it is pivotal for establishing a management system, identification system, and governance of smart city data.

The ongoing activity of ETSI ISG CIM is reflected in regular API releases, drawing on feedback from developers, end users and stakeholders. It is also worth noting that the standards also include a test suite and interoperability test descriptions, including implementation conformance statement and test purposes descriptions.

The global standards initiative oneM2M brings together all components in the IoT solution stack. It avoids reinvention in favour of reusing existing technology components and standards. oneM2M's architecture defines a common middleware technology in a horizontal layer between devices and communications networks and IoT applications. This standardizes links between connected devices, gateways, communications networks and cloud infrastructure. It allows developers to mix and match components from different vendors.

oneM2M publishes general-purpose standards that apply to all industry verticals. This ensures a high degree of re-use. It also means that vertical applications can interoperate with one another. This ability to work across application silos adds significant value and promotes innovation and makes it a very relevant family of standards for the development of smart cities services.

oneM2M has developed technical specifications for a **Common Service Layer** that can be readily embedded within various hardware and software and relied upon to connect the myriads of devices in the field with M2M application servers worldwide (Fig. 4.2). These standards cover:

- Use cases and requirements for a common set of service layer capabilities,
- Service layer aspects with high level and detailed service architecture, for an access independent view of end-to-end services,

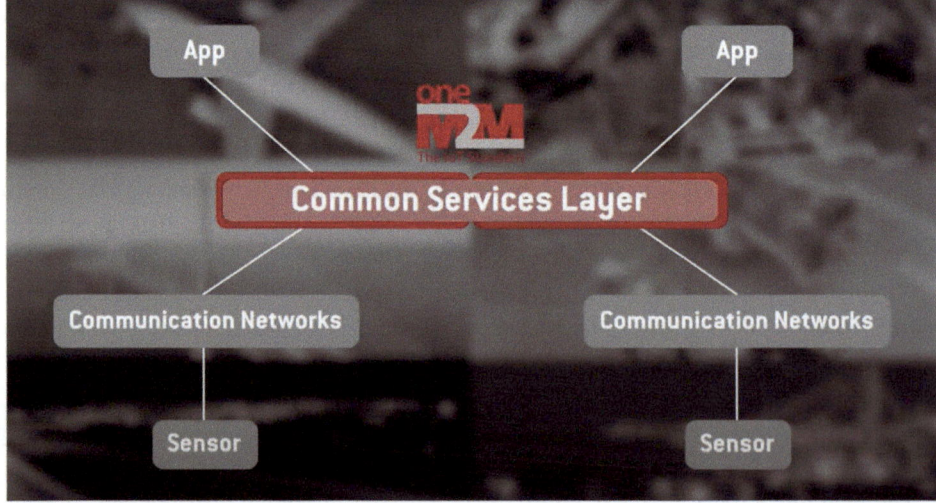

Fig. 4.2 Onem2m common services layer linking multiple communications networks with multiple domain applications (oneM2M 2022a, b)

- Protocols/APIs/standard objects based on this architecture (open interfaces and protocols),
- Security and privacy aspects (authentication, encryption, integrity verification);
- Reachability and discovery of applications,
- Interoperability, including test and conformance specifications,
- Collection of data for charging records (to be used for billing and statistical purposes),
- Identification and naming of devices and applications,
- Information models and data management (including store and subscribe/notify functionality),
- Management aspects (including remote management of entities),
- Common use cases, terminal/module aspects, including Service Layer interfaces/APIs between:
  - Application and Service Layers, and
  - Service Layer and communication functions.

By building upon well-proven protocols that allow applications across industry segments to communicate with each other, oneM2M enables service providers to combine different IoT devices, technologies and applications; a critical feature in their efforts to provide services across a range of industries. oneM2M has already been used in service provider deployments in the world and in Europe for smart city and smart mobility deployments.

The three technical bodies mentioned in this section coordinate their standardisation work with each other to avoid overlap and ensure consistency and complementarity.

4.3.5 Standards for Green and Sustainable Smart Cities

Smart cities IT managers need tools to monitor the deployment of sustainable smart cities services and the efficiency, including eco-efficiency and energy management, of their sites and networks. They also require means of implementing the most efficient broadband communications systems and physical networks.

(a) Energy efficiency and eco-design of ICT

ETSI Technical Committee on Environmental Engineering (TC EE) manages various engineering aspects of telecommunications and IT equipment in different types of installation.

More specifically, TC EE has published standards for life cycle analysis (LCA) and for the measurement and assessment methods of energy efficiency and power management of different types of ICT equipment, such as radio access networks, fixed networks and data centres. The reference standard for LCA is ETSI ES 203 199 V1.4.0 (2024-07) (ETSI 2024a, b, c).TC EE work also embraces innovative energy storage technologies for ICT equipment, for example to provide resilience in sustainable smart cities.

TC EE is also developing standards on eco-design requirements for servers and data storage products as well as mobile devices, that should have a positive impact on the development of refurbishment and recycling and reduce the amount of e-waste. Much of this work supports European Commission policies, regulation and legislation on eco-design matters.

TC EE interacts with ITU-T SG5 on the development of technically aligned standards supporting the reduction of ICT's impact on the environment and the development of circular economy. One example of standard contributing to develop circular economy is the work on digital product passports, starting with TS 103 811 v1.1.1 (2024-01). All these standards are fully relevant for use by smart cities and communities to reduce their environmental impact and achieve their green transition.

(b) Efficient (re)use of infrastructure

ETSI Technical Committee on Access, Terminals, Transmission and Multiplexing (TC ATTM) and particularly the working group on Sustainable Digital Multiservice Communities (ATTM SDMC) is working towards the creation, development and maintenance of standards relating to the sustainable deployment of ICT systems and implementation of services within cities and communities.

ETSI TS 110 174-2 (ETSI 2018b) standard considers multiservice networking infrastructure and associated street furniture. Its second part TS 110 174-2-2 (ETSI 2020) covers the use of lampposts for hosting sensing devices and 5G networking. Using lampposts and other street furniture for public wi-fi access or other city communications infrastructure was already done in pilot projects.

TC ATTM works closely with ETSI's Industry Specification Group on Operational energy Efficiency for Users (ISG OEU). These groups share the aim of improving the implementation of efficient ICT services, and increasing sustainability efficiency in operational networks and devices, as well as improving energy management efficiency and management of ICT waste in the operational period, from implementation until end of life.

(c) Key performance indicators

Using KPIs to measure and monitor improvements in smart cities will reinforce the impact of all initiatives and speed the achievement of vital goals such as the reduction of pollution and greenhouse gas emissions.

In TS 103 463-1 (ETSI 2017) and -2 (ETSI 2019b), a set of indicators for assessing smart city performance has been designed for ICT users, leveraging the definition of indicators from over forty existing indicator frameworks. The selected indicators focus on impact indicators, covering energy use, emissions of CO_2 and air pollutants, waste generation, and other people and prosperity themes, to respond to the priority needs of cities and citizens and support cities sustainability goals. The ICT users' indicators set described in this Technical Specification (TS) also includes KPIs to reflect the degree of smartness of a city. The indicators are classified using a mnemotechnic acronym "PPP" for People, Planet, Prosperity and complemented by a fourth overarching Governance category.

ETSI Industry Specification Group on Operational energy Efficiency for Users (ISG OEU) is supporting the development of standards for efficient sustainable communities, e.g., efficient engineering and global Key Performance Indicators (KPIs) for green smart cities, covering both residential and office environments.

By defining technical solutions for the implementation of resource-efficient infrastructures and services, and by specifying common KPIs for monitoring and tracking progress on their environmental impact, standards contribute to make smart cities and communities more sustainable.

4.3.6 Standards for Accessible and People—Centric Smart Cities

Standardisation also contributes to meeting citizen and consumer requirements, notably in terms of usability, accessibility, data security and safety.

(a) Citizen-centric services

Digital services should avoid being a series of upgrades to non-digital services, with resulting differences in approach and incompatibilities. When designing services, the citizens' use, and desired experience of city services should be the main elements to be considered. Convergence and integration of services may be required to respond to the citizens' needs in a seamless and user-friendly manner.

ETSI User Group (SC USER) has defined a user-centric approach in digital ecosystem in ETSI TR 103 438 (ETSI 2019a). With a "User Centric" approach, the users are above all elements: networks, applications or systems. They are at the centre of the architecture, and should be able to personalize their services, access them dynamically through the accesses offered during their actions, according to the QoS (Quality of Service) desired.

(b) Accessible community services

Concerning accessibility, the time when cities did not support independent living of people with disabilities or other accessibility needs should be soon behind us. Efforts have been made in the health sector for improving the quality of life and independent living of people with disabilities. Regulations have been adopted in the EU and worldwide on the accessibility requirements for products and services, applicable to smart city services.

Whilst there is gradual improvement in physical accessibility, for example in transport, building and street accessibility often requires improvement. Use of digital technology, for example for people to call for specialized transport facilities, needs to be made more widely available.

ETSI Technical Committee Human Factor (TC HF) has published a reference standard EN 301 549 (ETSI 2021) specifying the **accessibility requirements of ICT products and services**. This harmonized European Standard is under revision in support of the European Accessibility Act that will come into force in 2025. When the new version of the EN 301 549 will be published, all smart city services delivered over a digital user interface will be able to rely on this standard to ensure compliance with the legislation. This will encompass smart city services user interfaces such as any kiosks, ticketing or vending machines, public information displays, etc.

The accessibility requirements of the European Accessibility Act also apply to emergency calling, so that all people in need will be able to seek help in emergency situations.

(c) Safety and emergency communications

ETSI Technical Committee on Emergency Communications (TC EMTEL) is focused on the access to emergency services through different media, data transmission to public safety answering points, networks and IoT devices in emergency situations and in the context of the European Public Warning System. Its scope includes emergency communications between individuals and authorities/organizations, between authorities/organizations and between individuals.

Much of TC EMTEL's activity is centred on communications between IoT devices in emergency situations and on defining the technical specifications to enable the Next Generation 112 project that will provide network–independent **access to emergency services via a single European emergency number 112**, as defined in ETSI TS 103 479 V1.2.1 (2023-03) (ETSI 2023a, b) and related test specifications.

In support of the European Accessibility Act, TC EMTEL is developing a new harmonised standard on the accessibility and interoperability of emergency communications and for the answering of emergency communications by public safety answering points (PSAP).

TC EMTEL is also revising its technical specification TS 101 470 (ETSI 2013) of total conversation access to emergency services.

Many other standards are available, published by ETSI or other standardisation bodies, that contribute to the improvement of the citizens' experience of digital services, for instance covering data security, privacy and safety, that cannot all be cited here.

4.3.7 Future Standards Development

The success of standards is highly dependent on their adoption and implementation by industry and public authorities.

The ETSI standards mentioned in the previous Sects. 3.4, 3.5, and 3.6 have reached such high maturity level that ETSI Technical Bodies are maintaining and evolving them continuously considering new requirements and the feedback from developers, end users, regulators and other stakeholders.

In addition to the maintenance and evolution of existing standards, new topics are already under study for future standards development of relevance in the smart city context. These topics reflect the pace of innovation in information technologies. They include:

- building information modelling (BIM) standards to ensure interoperability of applications for connected buildings,
- the use of artificial intelligence technology in smart city applications and services, for example to anticipate and mitigate peaks and congestions,
- the specification of smart city local digital twins (LDT), to simulate and optimize the planning and deployment of services,
- the application of augmented and virtual reality in multiverse applications – i.e. mixing reality and virtual reality.

Given the fast-changing environment and buoyant ecosystem in which smart cities develop, many other innovation opportunities and standardisation needs may arise that are nearly impossible to predict.

4.4 Conclusions

This chapter introduced the most significant standardisation work relevant to smart cities conducted in ETSI, as well as a few illustrations of the potential for innovation in developing citizen-centric services and use cases.

Even if there appears to be many different standards applicable in the smart city context, each of them has a specific scope and function, such as defining a technical enabler (ontology, architecture, interface or protocol), a method for measuring KPIs or a way to comply with a given legislation. Standards are freely available to all and can be used to safeguard and improve the efficiency of smart city projects' implementations and the deployment of infrastructures and services.

If the reader only takes away one learning from this chapter, it should be the necessity to rely on global standards in smart cities ICT implementations and purchasing decisions, to achieve very significant potential benefits over time.

This chapter has also highlighted the importance of the involvement of stakeholders of the ecosystem in developing relevant standards, and it hopefully will trigger renewed engagement from city authorities, associations, solutions and service providers in standardisation activities.

Acknowledgements Many thanks to Patrick Guillemin, Technical Officer in ETSI for his sharing of knowledge of past developments of standards in ETSI, partner organizations and EU projects.

Special acknowledgement to Christophe Colinet, Smart City project manager of the City of Bordeaux, for his inspiration and enlightening testimonial about the role of standards in the deployment of smart cities and communities' projects.

Recognition should be given to the officials and rapporteurs of the oneM2M global partnership, ETSI TC SmartM2M, ISG CIM, TC EE, TC ATTM, TC HF, ISG OEU for producing high-quality standards in this domain.

I am also grateful to Anthony Brand, ETSI Chief Marketing Officer, and Issam Toufik, ETSI Chief Technical Officer, for reviewing this article before its publication.

All standards cited in the present chapter are available from the following websites:
SAREF Portal (etsi.org)
oneM2M Sets Standards for the Internet of Things and M2M
https://www.etsi.org/
For any enquiry about this chapter, please send your questions to info@esti.org

References

3GPP (1998) 3GPP: a global initiative (online). Retrieved Aug 12, 2024, from https://www.3gpp.org/about-us

Alliance for IoT and Edge Computing Innovation (AIOTI) (2015) Alliance for IoT and edge computing innovation (online). Retrieved Aug 12, 2024, from https://aioti.eu/

European Commission (2018) Enhancing synergies for disaster PRevention in the EurOpean Union (ESPREssO). Retrieved Aug 12, 2024, from https://cordis.europa.eu/project/id/700342/it

ETSI (2013). ETSI TS 101 470 V1.1.1: emergency communications (EMTEL); Total conversation access to emergency services. https://www.etsi.org/deliver/etsi_ts/101400_101499/101470/01.01.01_60/ts_101470v010101p.pdf

ETSI (2017). ETSI TS 103 463 V1.1.1: access, terminals, transmission and multiplexing (ATTM); Key performance indicators for sustainable digital multiservice cities. https://www.etsi.org/deliver/etsi_ts/103400_103499/103463/01.01.01_60/ts_103463v010101p.pdf

ETSI (2018a) ETSI TR 103 506 V1.1.1: SmartM2M; SAREF extension investigation; requirements for smart cities. https://www.etsi.org/deliver/etsi_tr/103500_103599/103506/01.01.01_60/tr_103506v010101p.pdf

ETSI (2018b) ETSI TS 110 174-2-1: access, terminals, transmission and multiplexing (ATTM); Sustainable digital multiservice cities; Broadband deployment and energy management; Part 2: Multiservice networking infrastructure and associated street furniture; Sub-part 1: General requirements. https://www.etsi.org/deliver/etsi_ts/110100_110199/1101740201/01.01.01_60/ts_1101740201v010101p.pdf

ETSI (2018c) Smart cities made simple. Video Retrieved Aug 12, 2024, from https://www.youtube.com/watch?v=pXSJmZcC2J8

ETSI (2019a) ETSI TR 103 438 V1.1.1: user centric approach in digital ecosystem. https://www.etsi.org/deliver/etsi_tr/103400_103499/103438/01.01.01_60/tr_103438v010101p.pdf

ETSI (2019b) ETSI TS 103 463-2: access, terminals, transmission and multiplexing (ATTM); Sustainable digital multiservice communities; Key performance indicators for sustainable digital multiservice areas; Part 2: Global KPIs for sustainable digital multiservice areas. https://www.etsi.org/deliver/etsi_ts/103400_103499/10346302/01.01.01_60/ts_10346302v010101p.pdf

ETSI (2020). ETSI TS 110 174-2-1: access, terminals, transmission and multiplexing (ATTM); Sustainable digital multiservice cities; broadband deployment and energy management; Part 2: Multiservice networking infrastructure and associated street furniture; Sub-part 2: The use of lamp-posts for hosting sensing devices and 5G networking. https://www.etsi.org/deliver/etsi_ts/110100_110199/1101740202/01.02.01_60/ts_1101740202v010201p.pdf

ETSI (2021). ETSI EN 301 549 V3.2.1: Accessibility requirements for ICT products and services. https://www.etsi.org/deliver/etsi_en/301500_301599/301549/03.02.01_60/en_301549v030201p.pdf

ETSI (2023a). ETSI GR CIM 002 V1.2.1: Context information management (CIM); Use cases (UC). https://www.etsi.org/deliver/etsi_gr/CIM/001_099/002/01.02.01_60/gr_CIM002v010201p.pdf

ETSI (2023b) ETSI TS 103 479 V1.2.1 (2023-03): emergency communications (EMTEL); Core elements for network independent access to emergency services. https://www.etsi.org/deliver/etsi_ts/103400_103499/103479/01.02.01_60/ts_103479v010201p.pdf

ETSI (2024a) Draft ETSI EN 303 760 V1.1.0: SmartM2M; SAREF guidelines for IoT semantic interoperability; develop, apply and evolve smart applications ontologies. Retrieved Aug 12, 2024, from https://www.etsi.org/deliver/etsi_en/303700_303799/303760/01.01.00_20/en_303760v010100a.pdf

ETSI (2024b) ETSI ES 203 199 V1.4.0 (2024-07): environmental engineering (EE); methodology for environmental life cycle assessment (LCA) of information and communication technology (ICT) goods, networks and services. Retrieved Aug 12, 2024 from https://www.etsi.org/deliver/etsi_es/203100_203199/203199/01.04.00_50/es_203199v010400m.pdf

ETSI (2024c) TS 103 811 v1.1.1 (2024-01): global digital sustainable product passport opportunities to achieve a circular economy. https://www.etsi.org/deliver/etsi_ts/103800_103899/103881/01.01.01_60/ts_103881v010101p.pdf

International Standards Organization (ISO) (2017) ISO/IEC 30182:2017—smart city concept model—guidance for establishing a model for data interoperability. Retrieved Aug 12, 2024, from https://www.iso.org/standard/53302.html

International Standards Organization (ISO) (2018) ISO 37120:2018—sustainable cities and communities—indicators for city services and quality of life. Retrieved Aug 12, 2024, from https://www.iso.org/standard/68498.html

oneM2M (2012) oneM2M: the IoT standard. Retrieved Aug 12, 2024, from https://www.onem2m.org/

oneM2M (2022a) oneM2M—the global standard for interoperable IoT systems. Retrieved Aug 12, 2024, from https://www.youtube.com/watch?v=5-gXAyYakJE

oneM2M (2022b) oneM2M—the global standard for interoperable IoT systems. Retrieved Aug 12, 2024, from https://www.youtube.com/watch?v=5-gXAyYakJE

SynchroniCity (2018) SynchroniCity—technology & life. Retrieved Aug 12, 2024, from https://www.synchronicity-iot.eu/

United Nations (UN) (2015) Transforming our world: the 2030 agenda for sustainable development. Retrieved Aug 12, 2024, from https://documents.un.org/doc/undoc/gen/n15/291/89/pdf/n1529189.pdf and https://sdgs.un.org/goals/goal11

Laure Pourcin is a Technical Officer at the European Telecommunications Standards Institute (ETSI), where she plays a crucial role in developing and maintaining global standards for telecommunications and information technologies. With a strong background in science and engineering and over 30-years' experience internationally in the telecommunications and IT industry, Laure is instrumental in coordinating technical committees, ensuring that standards meet the evolving needs of the industry. Her work supports innovation and interoperability across a range of technologies, including IoT, Environment Engineering, Cybersecurity, Human Factors and Accessibility. Committed to advancing digital transformation of society, Laure is recognized for her expertise and contributions to the Standardisation process within the tech industry.

Part IV
Smart City Standardization: National Efforts

United States Approach to Standards for Smart Cities

5

W. Michael Dunaway and Cheyney M. O'Fallon

Abstract

The Global Community Technology Consortium (GCTC) is a smart cities program established by the National Institute of Standards and Technology (NIST), the research laboratory of the U.S. Department of Commerce. Since its formation in 2014, the GCTC has built an international partnership of cities and communities, private-sector entities, and government agencies dedicated to improving urban and rural ecosystems through integration of advanced digital technologies. Smart city standards are being developed in collaboration with the NIST Standards Coordination Office and guidelines established by the U.S. National Technology Transfer and Advancement Act (NTTAA) and relevant federal directives. This chapter describes the U.S standards development approach for the diverse communities, cities, regions, and governing jurisdictions within the United States, and presents the NIST approach for defining standards and metrics for smart cities.

Keywords

Smart cities • Standards • Key performance indicators (KPIs) • Resilience • Sustainability

---

W. M. Dunaway (✉) · C. M. O'Fallon
Communications Technology Laboratory, National Institute of Standards and Technology, Gaithersburg, USA
e-mail: michael.dunaway@nist.gov

C. M. O'Fallon
e-mail: cheyney.ofallon@nist.gov

© The Author(s), under exclusive license to Springer Nature Switzerland AG 2025
L. Anthopoulos (ed.), *Smart City Standardization*, Synthesis Lectures on Computer Science, https://doi.org/10.1007/978-3-031-95959-2_5

5.1 Introduction

5.1.1 Origin of the NIST Global Community Technology Consortium (GCTC)

The Global Community Technology Consortium (GCTC) is a U.S. smart cities program established by the National Institute of Standards and Technology (NIST) within the U.S. Department of Commerce. The GCTC is a partnership of cities and communities, local and state government agencies, business enterprises, non-governmental organizations, universities, and research institutes dedicated to improving the urban environment through the integration of digital technologies. NIST coordinates both national and international dimensions of this partnership, in collaboration with other federal agencies and offices that sponsor smart city-related projects and research. [Throughout this chapter, the term "smart city" refers to any city, township, community, or region that adopts digital technologies with the goal of improving public services and operations and residents' quality of life.]

The GCTC originated from a 2013 NIST initiative called the "SmartAmerica Challenge," which had the goal of accelerating the development and adoption of advanced technologies to address a wide array of challenges facing cities, both in the U.S. and globally. The GCTC was formally established in 2014 as the Global City Teams Challenge and held its inaugural workshop in September 2014 at NIST's Gaithersburg, Maryland facility. The workshop included city teams, technologists, and researchers with experience in developing and integrating Cyber-Physical Systems (CPS) and Internet of Things (IoT)-based technologies at the municipal and regional levels. The principal challenge that the GCTC was established to address was articulated in a 2016 NIST Special Publication:

> Hundreds of cities and dozens of technology providers are working to realize civic benefits and potential profits across a broad range of services and markets. However, the critical goal of interoperability is in danger of being overwhelmed by the large wave of isolated and customized solutions, along with the accompanying proliferation of proposed standards and protocols. However, if too many details are standardized, innovation is overly constrained; if nothing is standardized, the result is non-interoperable clusters of function that are not easily integrated. (Global City Teams Challenge 2019).

The GCTC is based on the concept that a "smart city" is a community ecosystem in which advanced technologies are adopted in order to increase the efficiency, availability, and accessibility of city services to improve city operations, enhance public safety and community resilience, equitably distribute economic and social benefits, and improve overall quality of life for residents. The principal goal of the NIST program on Smart Cities is to support collaboration among innovative local governments and agencies, non-profits, private companies, and university research centers to overcome community challenges. In 2024, the program was redesignated the Global Community Technology Consortium to more accurately reflect the nature of the collaborative public private partnership that the GCTC has become.

5.1.2 Evolution of the GCTC and U.S. Smart City Movement

Beginning with its launch in 2014, the GCTC program evolved into a network of over 240 U.S. and international cities and communities, involving over 500 industry, academic, and government stakeholders who jointly develop and deploy advanced technologies for smart cities and communities. The program has helped create an ecosystem for information sharing in which communities can collaborate and exchange best practices to improve efficiencies, lower costs through economies of scale, and improve the lives and economic benefits to their populations. In 2016, the GCTC organized into working groups based on specific community services, infrastructure sectors, and mission areas. The organizational structure for the GCTC included twelve Technology Sectors depicted in Fig. 5.1. (Additional details and explanation of the GCTC history, organizational structure, and goals are contained in the NIST Special Publication, *Global Community Technology Challenge Strategic Plan 2024–2026*) (Dunaway et al. 2024).

This organizational structure provided a way of aligning the GCTC beyond a focus on projects within individual cities and communities. However, this structure reflected a characteristic seen in many cities wherein individual offices, agencies, and departments align themselves into relatively autonomous units that manage day-to-day operations, budgets,

Fig. 5.1 GCTC technology sectors (Dunaway et al. 2024)

and planning independent of each other. This separation or "stove-piping" of city departments or functions can lead to inefficiencies, duplication of effort, and a lack of internal transparency and awareness of existing intersections of technologies, data, and infrastructure. Consequently, the need for a unified vision and analytic framework for technology integration and deployment was identified as a challenge that GCTC cities would need to address in order to achieve the digital transformation of city infrastructure and services. Ultimately, the integration of digital technologies into city infrastructure is about making data available to community decision-makers and managers in order to improve the efficiency of city services and operations, while also making government more transparent and accessible to the public. That is, advanced technologies, IoT sensors and data sources are the means to an end—and not the goal—of smart city technologies and concepts. Over the last decade, smart city programs have evolved toward a broader understanding of what constitutes a "smart" city or community, and recent efforts within the GCTC emphasize the trustworthiness of technology, the validity of data, decision systems, and decision-making, and the need to define metrics and measures of success for the city as both an engineered system and as a social system. The term adopted in the GCTC for the ultimate goal of the smart city is "Integrity," as both an engineering concept (i.e., systems or design integrity) and as an ethical or moral construct reflecting the public's trust in technologies, data, decisions, and in their community decision-makers, leadership, and relationships. The "smart city" is now widely understood as a balance among technical, environmental, socio-economic, and cultural dimensions of the urban environment, and not simply the adoption of advanced technologies to increase efficiencies and cost-effectiveness. To that end, in 2021, the GCTC program initiated a focused research effort on defining the characteristics of trust, integrity, validity, and technological diversity as a foundation of the smart city that yields measurable outcomes in quality of life and overall community well-being. Figure 5.2 is a simplified depiction of this evolution within the GCTC, illustrating the increasing emphasis on the outcomes sought through integration of advanced technologies, and not on technology development and integration for its own sake.

5.1.3 Definition and Analytic Framework for a Smart City

In 2022, NIST published a foundational document on smart cities entitled *Framework for Key Performance Indicators for Smart Cities* (Serrano et al. 2022) (referred to as the "H-KPI Framework") that defines the smart community based on a holistic analysis of the intended outcomes of technology integration. The definition of "smart" adopted by NIST as the foundation of the "smart city" is as follows:

> Smart equals the efficient use of digital technologies to provide prioritized services and benefits to the community.

Fig. 5.2 Evolution of the smart city concept toward system and societal integrity as the objective of technology investment and digital transformation (Dunaway et al. 2024)

Within that definition, the parameters of a smart city are defined through evaluation along four dimensions:

- Number of digital services and benefits
- Efficiency in implementation, including re-use and dual-use
- Quality of services and benefits
- Alignment with community priorities

The NIST H-KPI Framework establishes an approach that addresses both quantitative measurement (number of services and benefits; calculated efficiency in implementation) and qualitative assessment (quality of services and benefits provided; alignment with community priorities) and provides community officials and leaders with a framework for assessing the outcomes and benefits of technology investments. The H-KPI Framework establishes a methodology for identifying, classifying, and evaluating system interactions at three levels of analysis: *Technologies, Services*, and *Benefits*, each of which represents specific technology or service platforms within the Smart City. Figure 5.3 provides a depiction of the H-KPI Framework structure (detailed in reference (Serrano et al. 2022)).

Level 1—Technologies. Level 1 identifies enabling technologies and cyber-physical systems and their core capabilities. These include sensors and actuators, networks, data systems, and computational hardware and software. The path represents the means of transmission via wired or wireless networks. The destination is the application, platform, data store, etc. receiving the transmitted data or information.

Level 2—Infrastructure and Services. Level 2 describes the infrastructure and services that enable a city to function as an integrated system. Infrastructures include communications, transportation, energy, water, and buildings. Examples of key city services include emergency response and law enforcement, waste management, electrical power generation and distribution, and community resources such as educational and medical facilities.

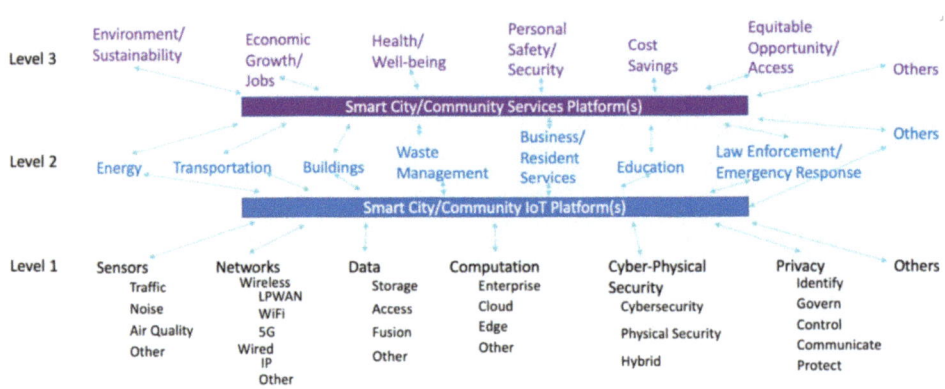

Fig. 5.3 Analytic structure of the NIST H-KPI framework for smart cities (Serrano et al. 2022)

Level 3—Community Benefits includes outcomes and applications that constitute the living and working environment that the service sectors establish for the community and its residents. These components and functions directly benefit people, organizations, and businesses and provide for safety and security, business and job growth, health care, environmental quality, and cultural enhancements. Performance metrics in Level 3 are human-centric and focus on factors such as quality of life, economic vitality and resilience, and public health and security, among others.

A key consideration for any smart city deployment is the role of interoperability in ensuring that technologies integrated into city infrastructure and operations are fully compatible with existing or legacy systems and will accommodate future improvements and new applications. For this reason, interoperability may be considered a fundamental requirement for any smart city project or program. The NIST smart city program has adopted the definition of interoperability from the NIST Framework for Smart Grid Interoperability Standards (NIST 2021) (Fig. 5.4), as a foundation for assessing technology investments as part of city digital transformation, as well as the foundation for local, regional, or national efforts in achieving standardization of smart city technologies, initiatives, or programs.

In 2025, the NIST Smart Cities program began to transition the management and community outreach to an affiliated non-profit organization, OpenCommons (www.opencommons.org), a long-standing member of the GCTC. Research in smart technologies, applications and systems continues to be conducted within the Smart Connected Systems Division of NIST, while OpenCommons focuses on building the smart cities "community" and developing an information sharing and analysis capability and repository for data, methodologies, lessons-learned, and applications of benefit to smart cities and communities. That work continues through the GCTC public private partnership that OpenCommons now leads.

> **"Interoperability" as defined in the**
> **NIST Framework and Roadmap for Smart Grid Interoperability Standards, Release 4.0**
>
> Interoperability [i]s the capability of two or more networks, systems, devices, applications, or components to work together, and to exchange and readily use information — securely, effectively, and with little or no inconvenience to the user.
>
> The benefits of interoperability are broad and reach all stakeholders at all scales. Interoperability is a hedge against technology obsolescence, maximizes the value of equipment investments by increasing usage for secondary purposes, and facilitates combinatorial innovation by allowing coordinated small actions across diverse stakeholders and devices to have grand impacts. The interoperability value proposition can be realized in any system domain, from the utility to the customer and beyond.
>
> Interoperability is therefore key to maximizing the benefits of technology investments. Yet because it is not easy to directly quantify the value of seamlessly exchanging a single bit of information in a complex system … the value of interoperability is most often thought of in the context of what is avoided: the expensive and time-consuming set of activities necessary for one-off integrations of incompatible systems.
>
> NIST Framework and Roadmap for Smart Grid Interoperability Standards, Release 4, 2021 pp. 1 – 3.

Fig. 5.4 Interoperability as a foundation for smart city development and standardization (NIST 2021)

5.2 Background

5.2.1 Standards Development in the U.S.

The National Institute of Standards and Technology (NIST) is the research institute of the U.S. Department of Commerce and lead authority on standards within the federal government. NIST coordinates policies and regulations to create an environment that reduces barriers to standards adoption for U.S. industry, while facilitating U.S. private and public sector engagement in international standards development organizations and activities. NIST's provenance on standards originates with the 1781 Articles of Confederation (precursor to the U.S. Constitution), which established that,

> The United States in Congress assembled shall also have the sole and exclusive right and power of regulating the alloy and value of coin struck by their own authority, or by that of the respective states—fixing the standards of weights and measures throughout the United States. (Articles of Confederation 1781)

From 1830 to 1901, management of this responsibility resided in the Office of Standards and Measures in the U.S. Coast and Geodetic Survey of the Department of the Treasury. From its earliest days, the concept of a federal office with oversight of standards development and management was seen as a balance (often, a competition) among interests between the need for manufacturing, construction and measurement compatibility and uniformity (along with tariff regulation), and the U.S. preference for a laissez-faire system of standards creation, adoption, and compliance exercised by private sector entities and state authority, while infringing minimally on individual rights and local jurisdictions.

In 1884, the chief of the Office of Weights and Measures, Charles Sanders Pierce ("a brilliant scientist, philosopher, and logician" (Cochrane 1974), and founder of the philosophical school of Pragmatism), stated that his Office "need not enter upon the business of inspecting commercial standards, because that is done already by the States in a satisfactory way." However, in light of rapidly advancing American capabilities in scientific research, industry, transportation, and communications—and at the request of many representing industry, manufacturing, and agriculture—the mission for standardizing weights and measures was established by a 1901 Act of Congress within the Department of Commerce and Labor as the National Bureau of Standards, the first physical science research laboratory of the federal government (Cochrane 1974).

The wisdom of establishing a central authority for standardization became tragically clear during the Great Fire of Baltimore, Maryland in February 1904, which destroyed 1526 residential and commercial buildings of the central city (Fig. 5.5) (Christhilf et al. 1904). A significant contributing cause of the catastrophe was the fact that responding firefighting units from surrounding jurisdictions discovered upon arrival that their firefighting hoses could not be connected to Baltimore city hydrants because of the incompatibility of hose couplings and lack of any adapter fittings that could have accommodated the differences in design. The lack of a national, or even regional, standard in simple emergency equipment contributed directly to the $150 million in damage to the city ($3.84 billion in 2014 dollars) and consequent unemployment of 35,000 residents. In the ensuing investigations, it was determined that, far from a local or regional problem, there were over 600 sizes and designs of firefighting couplings used in cities across the country (Schooley 2000; Seck and Evans 2004).

With that tragic event as backdrop, the National Bureau of Standards and its successor agency, the National Institute of Standards and Technology (renamed in 1988), have been at the forefront of both incremental and sometimes dramatic investigations and research in technology and disasters, such as the structural failure of the World Trade Centers on 9/11/2001; city-wide destruction from an EF-5 tornado in Joplin, Missouri, 2011; and the impact of Hurricane Maria in Puerto Rico, 2017 (NIST 2023a). In the current era, NIST conducts basic and applied research in measurement science and supports standards development and technological and scientific innovation. In collaboration with

Fig. 5.5 The Great Baltimore Fire February 7th, 1904. Photo: Christhilf Bros. & Litsinger, Publ., Balto Md 1904 (Christhilf et al. 1904)

private sector industry and business and national research institutions, NIST has led and catalyzed improvements in standards development and adoption within the U.S. and globally through participation in international Standards Development Organizations (SDOs). A concise history of NIST can be found in the NIST Online Archive article "NIST at 100: Foundations for Progress" (NIST 2023b).

5.2.2 Federal Standards, Frameworks, and Guidelines

Generally, NIST does not develop standards on its own. Rather, NIST produces frameworks and guidelines facilitating the efforts of U.S. standards development organizations (SDOs) and private enterprise to adopt standards to achieve efficiencies in their processes, products, and services. The NIST frameworks and guidelines serve to reduce the complexity and cost of innovation for businesses, big and small. In the United States, a sector-specific and decentralized system of standardization emerged after World War II, obviating the need for a centralized process managed by government agencies. In this regard, the U.S. approach to standards development differs from that of many other nations, economic entities, and SDOs, and reflects a strategy grounded in the strength, size, and diversity of the U.S. economy and institutions, and a historic preference for avoiding government intervention in matters that can be assumed or led by state authority or the private sector. The national approach to standards development is described in several key documents that explain the U.S. standards strategy and the specifics of how that strategy is to be implemented (Budget 2016; Devaux 2001; United States Congress 1996). Collectively, these documents establish a U.S. approach to standards development, and direct federal agencies to use private sector standards wherever possible in lieu of creating government-unique standards, while advancing commercialization through cooperative research and development agreements that make federal laboratory facilities available to the private sector.

Given the size and diversity of the U.S. economy, there is a robust collection of stakeholders with interest in efforts to enhance standardization. The lead coordinating organization is the American National Standards Institute (ANSI), founded in 1918 as a non-profit federation to enhance the competitiveness of U.S. industry and promote the development and adoption of voluntary consensus standards. Since its founding, ANSI has served as the central clearinghouse for member organizations, and issues guidelines for standards bodies, accredits U.S. SDOs, and publishes a United States Standards Strategy that "establishes a standardization framework built upon the traditional strengths of the U.S. system—consensus, openness, and transparency" (American National Standards Institute (ANSI) 2020). ANSI also serves as the U.S. representative to the International Electrotechnical Commission (IEC) and the International Organization for Standardization (ISO), in which NIST also participates. (The NIST Smart City program participates on the

U.S. Technical Advisory Group (US TAG) to the IEC System Committee and Joint IEC/ISO working groups for Smart Cities (International Electrotechnical Commission 2023)).

To summarize, standards development in the U.S. is guided by the over-arching principle that standards be developed through a transparent process that is "voluntary, consensus-based, and contains provisions for the non-discriminatory, royalty-free, or reasonably compensated availability of relevant intellectual property to all" (Devaux 2001). To reinforce this principle, federal agencies are required to use voluntary, consensus standards developed through private sector initiative rather than through government direction, except where impractical or inconsistent with law. Government use of voluntary consensus standards reduces federal costs for development and procurement and lowers compliance burdens, while aligning standards with national needs and encouraging economic growth and competition. This approach is consistent with the U.S. Government's commitment to the World Trade Organization's Technical Barriers to Trade Committee principles of transparency, openness, impartiality, and consensus (World Trade Organization 2023). While the complexity brought by this scale of effort and diversity of participants can seem antithetical to the efficiency of a top-down standards process, the U.S. approach of voluntary and consensus-based participation in standards development generates mutual interest and open competition among parties who have both vested interests in outcomes and the technical knowledge and capability to contribute meaningfully to the process.

5.2.3 National Standards Strategy for Critical and Emerging Technology

In May 2023, the National Security Council published the U.S. Government National Standards Strategy for Critical and Emerging Technology (USG NSSCET, hereinafter abbreviated as NSSCET), as an approach for U.S. federal agencies, business and industry, technology developers, and research institutes to partner on strategic goals, international engagements, scientific research, and lines of effort for managing the challenges of a new generation of critical and emerging technologies (NIST 2023c; The White House 2023). The NSSCET is intended to support and complement the ANSI U.S. Standards Strategy with a focus on critical and emerging technologies. The NSSCET reinforces the federal partnership with the private sector through four objectives: (1) investment in research and development; (2) participation in standards development; (3) workforce growth; and (4) integrity and inclusion as core principles of standards governance. These goals serve to address both national and international priorities for integrating new technologies into critical infrastructure, information technology systems, and society as a whole. The NSSCET identifies seven technology domains critical to U.S. strategic interests (Fig. 5.6) and are regularly reviewed and updated by the National Science and Technology Council (The White House 2022).

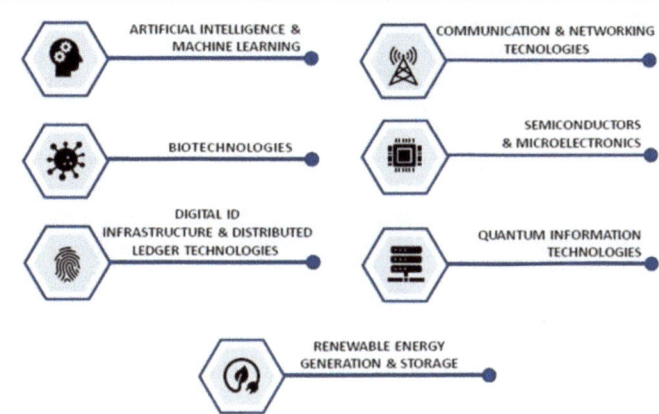

Fig. 5.6 Critical and emerging technologies identified in the USG NSSCET (The White House 2023)

NIST is charged with developing an implementation plan for the NSSCET, and in August 2023 established a dedicated project team aligned with the NIST Standards Coordination Office to lead the effort (NIST 2023d). Given the complexity and criticality of emerging technologies and the national scope of the effort, NIST initiated a program of collaboration and information gathering among agencies, private sector businesses, research institutions, and SDOs to define the best long-term approach for implementing the NSSCET. The rapid evolution of advanced technologies and their integration into society emphasizes the need for a comprehensive strategy for CET adoption, as exemplified by current NIST guidelines for safely integrating advanced technologies (e.g., the NIST Cybersecurity Framework 2.0 (NIST 2023e), and the NIST Artificial Intelligence Risk Management Framework (NIST 2023f)). Figure 5.7 illustrates a general process for developing the implementation plan.

As a starting point for this process, NIST issued a "Request for Information (RFI) on Implementation of USG NSSCET" (National Archives. Federal Register 2023), which solicited input to a series of questions relevant to CET from the business and industry sectors and the general public. NIST then launched a series of "listening sessions" in major cities across the nation to gain perspectives from U.S. business, industry, government agencies and the public in defining what the federal role should be in coordinating the development and integration of national standards for critical and emerging technologies. Among the questions asked, ten -in particular- highlight the relationship between the federal government and private sector partners in standards development (Fig. 5.8).

As the questions illustrate, the foundation of the U.S. standards development process is the partnership between the U.S. federal agencies and the nation's business, industry, and research sectors. This collaborative process for defining a national strategy for standards

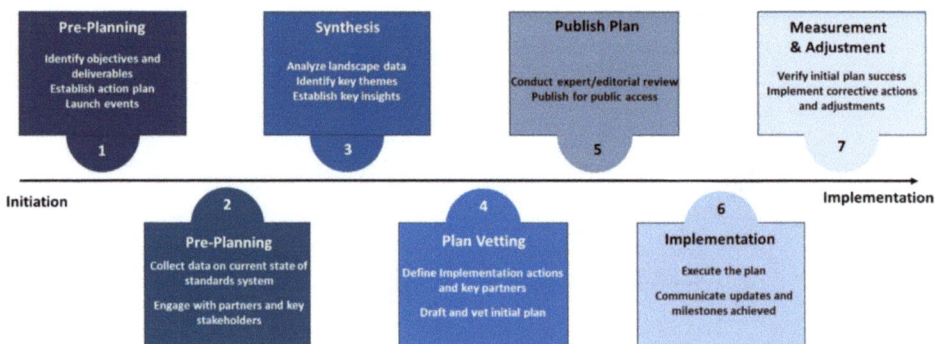

Fig. 5.7 NIST sequence for developing an implementation plan for the USG NSSCET (The White House 2023)

- What are the most important challenges faced by the private sector (i.e., industry, including start-ups and small- and medium-sized enterprises, academic community, and civil society organizations) when participating in standards development activities for CET? How can these challenges be addressed?
- How can the U.S. Government facilitate the adoption of standards-based CET by industry stakeholders, including start-ups and small- and medium-sized enterprises?
- How can the U.S. Government establish policies that promote standards development for CET as a component of U.S. innovation culture?
- How can the U.S. government better support publicly funded and private sector research in standards development activities for CET?
- How can the U.S. government increase the amount and consistency of private sector engagement in standards development activities for CET?
- How can the U.S. government foster early collaboration and private sector stakeholders to identify standards for CET that would encourage market and regulatory acceptance?
- How can the U.S. government work with private sector stakeholders to more effectively coordinate with international partners and reinforce private sector-led standards development?
- What standards development activities for CET can the U.S. Government and private sector stakeholders promote or develop to encourage increased participation by students and trainees?
- How can the U.S. Government work with international partners to ensure that standards for CET support free and fair market competition in which the best technologies come to market?
- How should the U.S. Government share information on standards development activities for CET with like-minded partners and allies?

Fig. 5.8 Representative questions from the USG NSSCET request for Information (National Archives. Federal Register 2023)

in critical and emerging technologies reflects an ongoing process directed at producing a report of findings, an implementation plan, and a collaborative process among government agencies and private sector entities. The outcome will reflect the U.S. approach for a federally chartered, public–private partnership of business, industry, and research institutions and will result in a higher degree of technical and commercial value for the nation, while establishing a methodology for future standards development for emerging technologies.

5.3 Research Approach for U.S. Smart City Standards

In addition to the seven technology domains illustrated in Fig. 5.6, the NSSCET identifies six areas of critical technology application that will have implications for the U.S. economy and national security. Figure 5.9 identifies the key technology applications defined in the NSSCET.

First among those domains is Automated and Connected Infrastructure—to include Smart Cities, Internet of Things and applications that rely on networked systems. In identifying smart cities as an area for technology development and standardization, the NSSCET presents the opportunity for U.S. SDOs, private sector entities, research institutions, cities, and communities to participate in the standards definition process for smart cities. However, the challenge of developing and implementing uniform standards for U.S. smart cities can be brought into perspective by reference to data from the U.S. Census Bureau, the statistical division of the Department of Commerce (National Academy of Public Administration 2023; Statista 2019; U.S. Census Bureau 2022).

Number of State and Local Governments in the United States

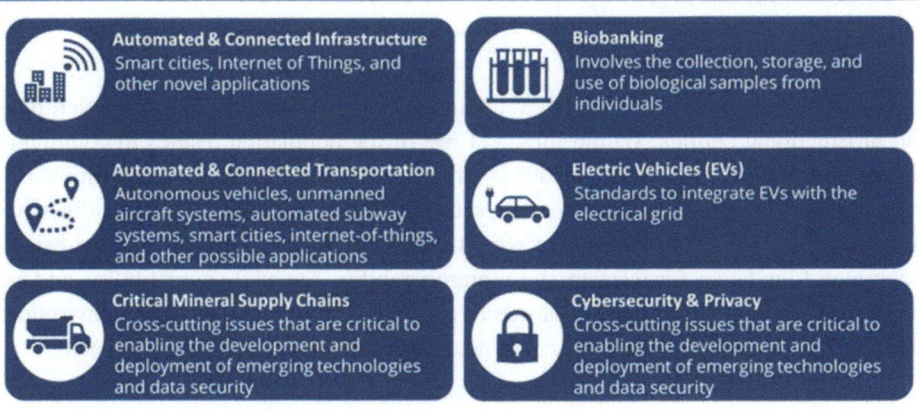

Fig. 5.9 Specific applications of critical and emerging technologies (The White House 2023)

State govts	50+ District of Columbia
County	3031
Municipal	19,495
Township	16,253
Special districts	38,542 (e.g., Fire, Water, Sewer, etc.)
School districts	12,754
Total Local Govt Entities	90,126

Number of cities, towns, and villages (incorporated places) in the United States

Population size	Number
1,000,000 or more	10
500,000 to 999,999	27
250,000 to 499,999	52
100,000 to 249,999	225
50,000 to 99,999	466
25,000 to 49,999	741
10.000 to 24,999	1521
Under 10,000	16,410
Total Incorporated places	19,452

Percent of U.S. Population in Urban Versus Rural Communities (2020 Census Estimate)

Urban centers (population greater than 50,000)	233,777,857	71.6%
Urban clusters (population of 2500 to 50,000)	29,588,545	9.1%
Rural (population less than 2500)	63,202,906	19.3%

To summarize, there are 19,452 U.S. communities operating 90,126 local governmental and jurisdictional entities. Within these, 71.6% of the population resides in communities greater than 50,000, and 28.4% resides in smaller cities, rural areas, and townships of less than 50,000. Therefore, defining a single set of standards appropriate for this diversity of communities—that is, standards that would gain nation-wide acceptability and establish a common foundation for infrastructure interoperability, regional compatibility, and community collaboration—is thus a particular challenge. In such a complex environment, defining U.S. smart cities standards will necessarily involve a two-part approach:

- The initial step will define or identify smart city standards that could be broadly applied within any city, community, and region, and would thus be "standardizable" to any aspiring smart city or community. This set of "extensible, scalable smart city standards" would be particularly relevant in building smart city collaborations across regions, within neighbourhoods and communities of larger metropolitan cities, and across rural areas with dispersed towns and populations.

- Of equal importance is identifying the set of smart city standards that would be more applicable to individual cities having unique characteristics as defined by, for example, geography and environment, exposure to specific hazards, demographic and socio-economic make-up, governance structure, culture and history, and other factors by which communities define themselves and celebrate their unique characteristics and identity.

The potential smart city standards defined by this analysis would yield one set of standards potentially applicable to all cities, and a second set of standards that may not be common to all communities but would nevertheless be valuable as a basis for regional collaboration and for information-sharing and development of best practices among communities having similar characteristics.

ANSI currently maintains a directory of 87 national and international standards for smart cities and communities and has published conceptual models, principles, requirements documents, and guides for metrics, along with a range of technology standards for city systems and applications that relate to the broad category of "Smart and Sustainable Cities" (ANSI 2023). To date, however, there has been no coordinated U.S. effort to define a unified body of standards for smart cities and communities that could inform research and development, municipal planning and design, local procurement processes, or community collaboration efforts, and thus simplify the process of implementation of advanced technologies. Reference by the NSSCET to smart cities and IoT technologies as applications of critical and emerging technologies presents the opportunity for developing standards for smart cities that address current, as well as future technologies.

5.3.1 Applying the NSSCET to U.S. Smart Cities

As an approach to developing a standards regime for U.S. smart cities, researchers from U.S. Smart Cities programs are collaborating with NIST research laboratories and the Standards Coordination Office to define requirements and standards to enable communities to capably and safely integrate CET into their digital transformation and smart city strategies. Within the GCTC program, communities can participate as pilot cities for testing digital technologies and concepts, with the goal of sharing knowledge and experiences to benefit the broader smart city ecosystem. At the same time, knowledge gained through case studies and applied research will enable the GCTC and OpenCommons to build a database and knowledge network of current smart city applications and emerging technologies (National Institute of Standards and Technology 2019).

- What are the most important challenges faced by cities, communities, and private sector partners in developing standards for smart cities? How can these challenges be addressed?
- How can the U.S. government facilitate the adoption of standards for smart cities and communities in partnership with community stakeholders and local governments?
- How can the U.S. government promote standards development in smart cities and communities as a critical component of U.S. culture of innovation?
- How can federal agencies better support publicly funded and private sector led research in standards development and adoption in smart cities and communities?
- How can federal agencies encourage the amount and consistency of private sector engagement in standards development in smart Cities and communities?
- How can the U.S. government foster collaboration with private sector stakeholders to define standards for smart city technologies that would encourage market and regulatory acceptance?
- How can federal agencies and private sector partners encourage participation in standards development activities by students and trainees?
- How can government work with business, industry, and international partners to reinforce private sector-led standards development in Smart Cities and related technologies?
- How should the U.S. government best share information on standards development activities for Smart Cities and Communities with like-minded international partners and allies?

Fig. 5.10 Questions to frame the application of the USG NSSCET to smart cities

Beginning in 2024, the NIST Smart City program began an active collaboration with the NSSCET Implementation Team, ANSI, and GCTC member communities in a series of workshops and community collaborations to define the best approach for designing and implementing standards for smart cities. As a starting point, the NSSCET RFI survey was modified to serve as the baseline for discussions between the NIST Smart Cities program, GCTC member communities, and private sector partners, research institutes, and national SDOs in defining U.S. standards for smart cities. Figure 5.10 provides some representative questions developed to guide that process.

This coordinated approach with the NSSCET will enable the GCTC Smart City program to capture community and private sector perceptions for adopting smart city standards, while reflecting the goals of the NSSCET in prioritizing smart cities as an application of Critical and Emerging Technologies.

5.3.2 Case Studies in Smart Cities

The process for defining smart city standards will parallel the method and strategy established by the NIST NSSCET Implementation Team and will be defined over the course of

Fig. 5.11 Holistic City KPIs assessment of spatial design factors. © State of Place (Built Environment Data—State of Place). Used with permission (Alfonzo 2023)

the next several years (Fig. 5.11). However, there are efforts currently in progress within the GCTC that address fundamental principles for smart city development that—while not yet establishing specific technology standards—will shape future technology adoption and standards development in U.S. smart cities. Four examples of efforts within GCTC cities are currently establishing standards that may be extensible to any community, city, or region.

5.3.2.1 Holistic Key Performance Indicators for Smart Cities

Several cities in the GCTC have begun applying the Holistic KPI Framework (Serrano et al. 2022) for analyzing the network of relationships, data, sensors, city services, and community outcomes in their smart city programs. The initial focus in several of the pilot cities is on improving public safety, community resilience, and disaster recovery planning, and improving coordination between first responders, emergency managers, and city leadership, both elected and unofficial. This effort is a high priority within the GCTC program, given the numerous challenges faced by communities in public safety and health, and the impacts of extreme weather and environmental change that every community faces. Figure 5.12 depicts a current GCTC research effort on Holistic KPIs to assess community technology and environmental and spatial design factors that reinforce community goals in livability and sustainability (Alfonzo 2023). The picture highlights a number of the over 125 planning factors that enhance overall community resilience and cohesion and contribute to a smart, safe, equitable, and livable city.

Fig. 5.12 City dashboard showing digital twin display for the City of Coral Gables, Florida. © 2022 City of Coral Gables, Florida. Used with permission (City of Coral Gables 2023)

5.3.2.2 Trust, Integrity, Diversity, and Equity

An important area of research within the GCTC is directed at understanding the nexus between community trust, integrity, social cohesiveness, and the economic and social benefits that can be achieved through the adoption of advanced technologies in communities. To that end, the GCTC Trust, Integrity, Diversity, and Equity (TIDE) project focuses on identifying KPIs to assess the intersections of smart city technologies and applications with the outcomes sought to build community integrity at both the system and societal levels. The TIDE concept as applied within the GCTC program is not directed solely at measuring social and economic outcomes, but rather at defining a broad metric for understanding the holistic relationship between the integration of advanced technologies with the goals of the community. In this regard, "diversity" is more aligned with the broader definition established by the International Telecommunications Union (ITU) Standardization Sector's Study Group 20 on Internet of Things (IoT) and Smart Cities and Communities (SC&C) (International Telecommunications Union 2023) and characterizes diversity from five perspectives:

- Diversity of services and applications.
- Diversity of data and data sources.

- Diversity of devices, networks, and transmission methods.
- Diversity of sensing modes and sensor systems.
- Diversity of management and governance modes.

Considered in the context of the 19,452 incorporated cities and communities within the United States, this broader understanding of diversity has relevance along technological, operational and governance dimensions, as well as for the more commonly understood meaning of demographic and socio-economic aspects of a community. The long-range goal of this effort is to develop a methodology for measuring (directly or indirectly) the impact of programs within a city or community to improving overall quality of life and well-being and ensure the equitable distribution of benefits across the diverse communities that characterize a complex, modern city.

5.3.2.3 Open Source/Open Standards/Open Access

One of the goals of the GCTC program is that smart city technologies be built on an open source and open standards framework that encourages innovation and entrepreneurialism, maximizes investment opportunities, and ensures system compatibility among smart technologies, while enabling universal access to the knowledge, benefits, and opportunities for growth and workforce development that smart cities promise. Currently under investigation by several GCTC communities and organizations is the establishment of an Open Source Program Office (OSPO) as a catalyst for community and regional collaboration on application development. An OSPO can streamline research and development of smart city applications and build a unified strategy to ensure that individual cities and agencies do not have to navigate the software space on their own, which is often a challenge for communities with limited technical staff, expertise, or resources. Additionally, the OSPO can assure equal opportunities for participation by smaller developers and companies in city infrastructure improvement, guard against service provider lock-in through acquisition of proprietary technologies, and future-proof community investments against obsolescence. The OSPO offers potential for both technological innovation and for devising business models that enable developers to share technology applications across communities while maintaining their intellectual property rights.

5.3.2.4 Human-Centered Design for Public Data Display

A relatively mature smart city concept is the adoption of internet-based city dashboards and websites to improve public access to information on city services, operations, and public service notifications. This technology application has become common throughout smart cities, including many smaller jurisdictions (provided that the community has the technical workforce to develop and manage the IT infrastructure). Several GCTC cities have begun developing city displays that provide publicly accessible 3-dimensional representations of the community (digital twins) that integrate with city agencies and services to provide on-demand community information. Figure 5.12 provides a view of the integrated city dashboard of the City of Coral Gables, Florida, which provides public service information in a user-friendly 3-dimensional mapping system and displays real-time information of city conditions and systems (City of Coral Gables 2023). A goal of this effort is to incorporate principles of human centered design into the development of this resource, so that it is accessible, intuitive, visually compelling, and user-friendly, thus enhancing transparency of government operations and services, and reinforcing trust within the community. Examples of successful implementations, such as that in Coral Gables, will guide GCTC efforts in research and standards development for publicly accessible data information display systems, and build a knowledge base of resources and applications with emphasis on human-centered design through collaboration with communities, agencies, and developers.

5.4 Conclusions: Toward a Framework for Smart City Standards

The design and implementation of a U.S. standards regime for smart cities, like other standards development efforts, will evolve as fundamental concepts, technologies, community priorities, and threats to community resilience and sustainability evolve. The research effort being led by the NIST Smart Connected Systems Division will investigate standards for existing smart city technologies and processes, while coordinating with the NIST Standards Coordination Office, ANSI, and other U.S. SDOs to define standards, methodologies, and protocols for incorporating smart city technologies that are currently available, as well as those that will emerge in the future. Concurrently, NIST membership on the U.S. Technical Advisory Group (US TAG) Systems Committee, and technical committees for smart cities in IEC and ISO, smart cities working groups of the Institute of Electrical and Electronics Engineers (IEEE), and International Committee for Information Technology Standards (INCITS) and other international SDOs will enable U.S. smart cities programs to gain insights from international sustainable smart cities and communities, and where appropriate, enable compatibility with international standards regimes in advanced technologies. In the U.S., NIST will conduct research and development efforts in collaboration with GCTC member cities and communities, technology developers, businesses and industries, and research institutes to create a "Framework for Smart City Standards" that (1) reflects a U.S. standards development process grounded on "voluntary, consensus-based, and private sector led development;" (2) is representative of the diverse interests and character of U.S. cities, communities, and jurisdictions; and (3) is extensible, scalable, and applicable to any community, city or region in the U.S.

The fundamental question to be answered in developing a standards regime for smart cities is "What smart city concepts, technologies, and frameworks could be universally applied across all cities and are thus amenable to being adopted as standards; and which technologies and approaches might not be universally applicable, owing to the diversity of cities, towns, communities, and jurisdictions in the United States?" Defining smart city technology standards that are extensible, scalable, and applicable to the 19,452 cities and communities that make up the United States will remain an enduring challenge. To address this challenge, four principles that guide NIST research in smart cities and the GCTC program will form the basis for defining a U.S. Framework for Smart City Standards:

- Establishment of standard metrics and protocols based on holistic key performance indicators that are measurable, scalable, appropriate to community needs, and transparent to community residents.
- Development of smart city technologies and applications based—to the degree feasible—on open source, open standards, and open access protocols and implementation strategies.
- Design and incorporation of Trust, Integrity, Diversity, and Equity (TIDE) protocols into smart city programs to establish a more holistic approach to community digital transformation that ensures community benefits to all of a community's residents and neighborhoods.
- Integrating principles of human-centered design for analyzing and displaying complex city data to make information more usable, intuitive, and actionable for community leaders and decision-makers, and more accessible and meaningful to residents and the general public.

With these four principles as a foundation, the future mission of NIST smart cities research endeavor, and the GCTC and OpenCommons information sharing structure is to define standards and metrics for smart cities that can streamline the process that U.S. cities, communities, and regions adopt in pursuit of digital transformation and achievement of their smart city goals.

5.5 Review Questions

Q1: The National Institute of Standards and Technology established a smart city program in 2014 as the Global City Teams Challenge (GCTC) with the specific goal to solve a problem that was emerging with the adoption of digital technologies by cities and communities. What was the specific problem or challenge the GCTC program was intended to address?

A1: With the proliferation of U.S. smart city initiatives and solutions, the critical goal of interoperability among city systems was being overwhelmed by the adoption of isolated and customized solutions, along with an accompanying proliferation of proposed standards and protocols. The challenge that the GCTC was intended to address was to provide a basis for standardization of city digital transformation and establish a venue for information sharing without stifling local innovation and entrepreneurism.

Q2: In developing a framework for smart city key performance indicators (KPIs) NIST developed a concise definition of "smart" as applied to a smart city. What is that definition and the four criteria by which "smartness" is understood?
A2: The definition of "smart" adopted by NIST as the foundation of the "smart city" is
> "the efficient use of digital technologies to provide prioritized services and benefits to the community."

The parameters of a smart city are measured and assessed along four dimensions:
- Number of digital services and benefits
- Efficiency in implementation, including re-use and dual-use
- Quality of services and benefits
- Alignment with community priorities

Q3: What are some of the reasons for the U.S. preference for a laissez-faire approach to standards development?
A3: The U.S. approach to standards development reflects a strategy grounded in the strength, size, and diversity of the U.S. economy and institutions, and a historic preference for avoiding government intervention in matters that can be assumed or led by state authority or the private sector. The U.S. approach to standards development seeks to capitalize on the strengths inherent in a pluralistic society with minimal infringement upon the individual rights of local jurisdictions, especially in technical matters. Such an approach also offers maximum latitude for the emergence of the best possible standards and accounts for potentially rapid changes in the technological state of the art. Finally, the U.S. approach leverages the strengths that make U.S. communities socially, culturally, and economically vibrant.

Q4: A central challenge for developing smart city standards in the United States is in designing a standards regime that can address the specific characteristics, priorities, and technological requirements of the 19,000+cities, towns, villages, and rural communities that comprise the U.S. NIST envisions a two-phase approach to smart city standards development to address the diversity of U.S. communities. What is the rationale and characteristics of this approach?
A4: The initial phase will define or identify smart city standards that could be broadly applied within any city, community, and region, and would thus be "standardizable" to any aspiring smart city or community. These standards would be particularly relevant in building smart city collaborations across regions.

The second phase will define a set of smart city standards that would be more applicable to individual cities having unique characteristics as defined by geography and environment, exposure to specific hazards, demographic and socio-economic make-up, governance structure, culture and history, and other factors.

The potential smart city standards defined by this two-part analysis would yield one set of standards potentially applicable to all cities, and a second set of standards that may not be common to all communities but would nevertheless be a valuable resource and a basis for regional collaboration and information-sharing and development of best practices among communities having similar characteristics.

Q5: What is the core value proposition of the Open Source Program Office (OSPO) or similar establishments within the smart city domain?
A5: OSPOs encourage innovation and entrepreneurialism, maximize investment opportunities, and ensure system compatibility among smart technologies, while enabling universal access to the knowledge, benefits, and opportunities for growth and workforce development. Adoption of an OSPO for coordination of smart city application development and implementation can also reduce redundant work across city bureaus or functions by creating a shared set of resources and capabilities.

References

Alfonzo M (2023) State of place. Data and AI to foster more equitable communities. https://www.stateofplace.co/. Accessed 20 Nov 2023

American National Standards Institute (ANSI) (2020) United States standards strategy (ansi.org)

American National Standards Institute. ANSI store catalog of smart cities standards. https://webstore.ansi.org/search/find?in=1&st=smart+cities. Smart and sustainable cities (ansi.org). Accessed 10 Nov 2023

Articles of confederation of 1781, Article IX, Paragraph 4. Cited from Wikipedia National Institute of Standards and Technology—Wikipedia. Accessed 21 Oct 2023

Budget OOMA (2016) OMB Circular A-119: federal participation in the development and use of voluntary consensus standards and in conformity assessment activities. Circular-119-1.pdf (whitehouse.gov)

Christhilf GE, Litsinger AL, Christhilf JC (1904) Photographic views and description of the Great Baltimore Fire. Christhilf Bros. & Litsinger, Publ., Baltimore

City of Coral Gables. Innovation and technology department. Digital twin smart city platform. Website. Coral Gables Smart City Hub (arcgis.com). Accessed 20 Nov 2023

Cochrane R (1974) Measures for Progress, National Bureau of Standards (NIST), p. 84. Found in NIST Digital Archives NIST Digital Archives—NIST Digital Archives (oclc.org)

Devaux C (2001) A guide to documentary standards. National Institute of Standards and Technology. Gaithersburg, MD. GOVPUB-C13-PURL-gpo29920.pdf (govinfo.gov)

Dunaway M, Roth T, Griffor E, Wollman D (2024) Global community technology challenge strategic plan 2024–2026. National Institute of Standards and Technology. SP 1900-02. https://doi.org/10.6028/NIST.SP.1900-02

Global City Teams Challenge. National Institute of Standards and Technology. Cyber-Physical Systems web page. https://www.nist.gov/ctl/smart-connected-systems-division/iot-devices-and-infrastructure-group/smart-americaglobal-0

International Electrotechnical Commission. Systems committee on smart cities: electrotechnical aspects of smart cities. Website. IEC—SyC smart cities: electrotechnical aspects of Smart Cities > Scope Accessed 02 Nov 2023

International Telecommunications Union. ITU-T SG20, Internet of Things (IoT) and Smart Cities and Communities (SC&C). ITU-T SG20: Internet of things (IoT) and smart cities and communities (SC&C). Accessed 18 Nov 2023

National Academy of Public Administration. Federalism US. number of government jurisdictions. Website. 1.1 Number of government jurisdictions—FEDERALISM.US. Accessed 02 Nov 2023

National Archives. Federal register. Request for information on implementation of the United States government national standards strategy for critical and emerging technology (USG NSSCET). https://www.federalregister.gov/documents/2023/09/07/2023-19245/request-for-information-on-implementation-of-the-united-states-government-national-standards. Accessed 02 Nov 2023

National Institute of Standards and Technology. NIST smart cities and communities framework series. Webpage. NIST smart cities and communities framework series | NIST

NIST framework and roadmap for smart grid interoperability standards, Release 4.0, NIST Special Publication 1108r4 (2021). https://nvlpubs.nist.gov/nistpubs/SpecialPublications/NIST.SP.1108r4.pdf

National Institute of Standards and Technology. NIST at 100: foundations for progress. NIST online archive. https://www.nist.gov/pao/nist-100-foundations-progress. Website Accessed 20 Oct 2023

National Institute of Standards and Technology. NIST disaster and failure studies disaster & failure studies | NIST. Website Accessed 01 Nov 2023

National Institute of Standards and Technology. U.S. government national standards strategy for critical and emerging technology. NIST program office website. U.S. Government National Standards Strategy | NIST. Accessed 02 Nov 2023

National Institute for Standards and Technology. Standards coordination office. Website: U.S. government national standards strategy | NIST Accessed 02 Nov 2023

National Institute of Standards and Technology (2023e) NIST cybersecurity framework 2.0. Computer security resource center website. CSWP 29, The NIST cybersecurity framework 2.0 | CSRC. Accessed 02 Nov 2023

National Institute of Standards and Technology (2023f) Artificial intelligence risk management framework. 2023. Program website. AI risk management framework | NIST. Accessed 02 Nov 2023

Schooley J (2000) Responding to national needs: the National Bureau of Standards becomes the National Institute of Standards and Technology 1969–1993

Seck M, Evans D (2004) Major U.S. cities using national standard fire hydrants, one century after the Great Baltimore Fire. NIST Interagency/Internal Report (NISTIR 7158), National Institute of Standards and Technology, Gaithersburg, MD

Serrano M, Griffor E, Wollman D, Dunaway M, Burns M, Rhee S, Greer C (2022) Smart cities and communities: a key performance indicators framework. National Institute of Standards and Technology. SP 1900-206. https://doi.org/10.6028/NIST.SP.1900-206.upd1

Statista. Number of cities, towns, and villages in the United States in 2019 by population size. Number of U.S. cities, towns, villages by population size 2019 | Statista. Accessed 02 Nov 2023

The White House (2022) Critical and emerging technologies list update. Report to the National Science and Technology Council. Critical and emerging technologies list update (whitehouse.gov)

The White House (2023) United States government national standards strategy for critical and emerging technology. Fact Sheet: Biden-Harris administration announces national standards strategy for critical and emerging technology | The White House. Accessed 02 Nov 2023

U.S. Census Bureau Statistics from National Academy of Public Administration. Found in National Academy of Public Administration 2022 Statistics from 2017 U.S. Census Bureau, FEDERALISM.US—America's challenges, intergovernmental solutions

United States Congress (1996) National technology transfer and advancement act of 1995. Public Law 104, 113. Key federal law and policy documents: NTTAA & OMB A-119 | NIST

World Trade Organization. Agreement on technical barriers to trade. WTO | legal texts—Marrakesh Agreement. Accessed 01 Nov 2023

Michael Dunaway is a Research Associate and former Director of Innovation in the Smart Connected Systems Division at the National Institute of Standards and Technology (NIST) and program lead for the Global Community Technology Consortium (GCTC), a U.S. Smart City program. Dr. Dunaway has led research programs in smart cities, community resilience and disaster preparedness at the University of Cincinnati and University of Louisiana at Lafayette, where he served concurrently as Director of the Louisiana Business Emergency Operations Center. Earlier assignments included positions at the National Headquarters of the American Red Cross; the Science & Technology Directorate, U.S. Department of Homeland Security; and the Cognitive, Neural, and Social Science Division of the Office of Naval Research. A graduate of the United States Naval Academy, he had previously served as a commissioned officer in the United States Navy, retiring at the rank of Captain. He holds a Masters degree from the Fletcher School of Law and Diplomacy at Tufts University, and a Doctorate in Systems Engineering from the George Washington University.

Cheyney O'Fallon is an economist and Research Associate with the Communications Technology Laboratory at the National Institute of Standards and Technology (NIST). He joined NIST in 2016 after earning his Ph.D. from the University of California, Santa Cruz, where his research focused on the competitive effects of natural disasters on electricity markets. He earned his BS degree in Economics from the University of Oregon in 2010. At NIST, Cheyney worked in the Applied Economics Office of the Engineering Laboratory before joining the NIST Smart Grid Program. He is a co-author of the NIST framework and roadmap for smart grid interoperability standards, release 4.0.

Cheyney's research has centered on applied econometrics and economic modeling of interoperable smart grid infrastructure. His research portfolio has included the varied topics of community resilience, life-cycle assessment relating to sustainable building design, as well as patterns in technology development and diffusion. More recently, Cheyney's research has focused on performance measurement and management in public and infrastructure sectors. Where possible and appropriate his research has prioritized the use of public data and open-source tools.

Smart City Standards in Canada (SCSC) 6

Val Wise

Abstract

This chapter explores smart city standardization in Canada. It delves into the national standardization effort, examining its history, development stages, and impact on smart city definition, implementation, and challenges. Readers will gain insights into strategic, process, and technical standards, along with valuable learning objectives for navigating smart city development.

Keywords

Smart cities • Standardization • Canada • Urban development • Technology smart city development • SCSC

6.1 Introduction

In the dawn of the twenty-first century, urban centers have become the lifeblood of economic, cultural, and technological advancement. As the global population increasingly gravitates towards cities, the concept of 'Smart Cities' has emerged as a beacon of sustainable urban development. In Canada, a country known for its vast landscapes and diverse cultures, the integration of smart city initiatives is redefining the urban experience.

Smart cities harness the power of advanced technologies and data analytics to optimize city functions, enhance public services, and foster economic growth. They are

V. Wise (✉)
Smart City Standards, Calgary, Canada
e-mail: wisevaleriy@gmail.com

U4CCS, Executive & Global Marketing, Calgary, Canada

characterized by their interconnected infrastructure, efficient transportation systems, and responsive governance. In the Canadian context, these cities are not just a vision of the future but are rapidly becoming a present-day reality.

Standardization plays a pivotal role in this transformation. It ensures interoperability between systems and lays down a framework for innovation and security. As Canadian cities adopt smart technologies, standardization provides the common language that enables diverse systems to work in harmony.

Urban development in Canada is uniquely challenged by the country's vast geography and varied climate. Smart city solutions offer a way to address these challenges head-on, using technology as a tool to bridge distances, connect communities, and create resilient infrastructures that withstand the test of time and nature.

> In 2014, a report titled 'Smart Cities: The Future of Urban Development' by Infrastructure Canada detailed various smart city initiatives across the country (Government of Canada 2014, p. 15).

ISO was tasked with developing interoperability standards to help municipalities coordinate and share best practices. This built upon international standards but was tailored to Canada's system of municipal governance.

Canada's smart city standards were developed by ISO (2017) to address the lack of coordination between municipalities pursuing independent smart city projects. ISO brought together experts from cities, industry and academia to establish a framework for ensuring interoperability. This chapter will outline the evolution and key components of Canada's smart city standards framework. The learning objectives are to understand how standards guide strategic planning, project management and technical implementation of smart initiatives. Canada's smart city standards were developed by ISO starting in 2015 to address the lack of coordination between municipalities pursuing independent smart city projects. ISO brought together experts from cities, industry and academia to establish a framework for ensuring interoperability.

This chapter aims to explore the intersection of technology and urbanization in the Canadian landscape. It delves into the standardization efforts that make smart cities possible, the innovative technologies driving change, and the impact of these developments on society and the environment. As we embark on this journey, we invite readers to envision the cities of tomorrow and the role Canada will play in shaping the urban landscapes of the future.

In the end of this chapter, the reader will be able to:

- Define smart city standards and their different levels (strategic, process, technical).
- Identify key standardization bodies and stakeholders involved in Canada.
- Understand the challenges and benefits of implementing smart city standards.
- Analyze the impact of standards on smart city development.

6.2 Background

This section explores the factors that influenced its development, such as national strategies, international initiatives, or city-led efforts. Additionally, the theoretical framework is discussed, including relevant technological advancements, business models, and legal considerations. The chapter will also delve into the organizational structures employed, such as technical committees and focus groups. Specific topics addressed by standards, like frameworks and technical specifications for areas like intelligent transportation or smart health, will be defined. There isn't a single, overarching standard for smart cities in Canada. Instead, development is influenced by a mix of factors:

6.2.1 National Strategy and International Initiatives

Canada hasn't established a single national smart city strategy. However, Infrastructure Canada's Smart Cities Challenge (2019) encouraged communities to develop smart solutions for local issues. This focus on local needs is a key aspect of the Canadian approach.

Internationally, Canada participates in efforts like the ISO 37120 (Sustainable cities and communities—Indicators for city services and quality of life) standard for indicators of sustainable and resilient cities. This standard provides a common language for cities and identifies key performance indicators (KPIs) for various aspects of urban life.

6.2.2 Theoretical Context

Canadian smart city initiatives often focus on:

- Technological artifacts: Sensors, data collection platforms, and connected infrastructure that gather and analyze urban data.
- Business models: Public–private partnerships and innovative financing mechanisms to fund smart city projects.
- Legal frameworks: Addressing data privacy, cybersecurity, and ownership of collected information in the smart city context.

6.2.3 Organizational Schemas

Several key organizations influence smart city development in Canada:

Standards Council of Canada (SCC): The Standards Council of Canada (SCC) is a federal Crown corporation that plays a pivotal role in promoting efficient and effective standardization in Canada. Established in 1970, the SCC is responsible for overseeing the

National Standards System, which comprises numerous organizations and experts dedicated to developing, promoting, and implementing national and international standards. The SCC's mission is to enhance Canada's economic performance and social well-being through the advancement of standardization activities.

As the central coordinating body for standards development in Canada, the SCC accredits standards development organizations (SDOs) and ensures that Canadian standards are aligned with global benchmarks. This accreditation process ensures that the standards produced are of high quality, relevant, and developed through a consensus-based process that involves a broad range of stakeholders, including industry representatives, government agencies, and consumer groups.

Furthermore, the SCC actively represents Canada's interests in international standardization forums such as the International Organization for Standardization (ISO) and the International Electrotechnical Commission (IEC). By participating in these global efforts, the SCC helps to ensure that Canadian products and services can compete effectively in international markets, fostering trade and innovation.

The SCC also plays a crucial role in enhancing the quality of life for Canadians by promoting standards that address health, safety, and environmental concerns. Through its work, the SCC contributes to the development of standards that support sustainable practices, improve public safety, and protect the environment.

In summary, the Standards Council of Canada is a key institution in the national and international standardization landscape, dedicated to advancing Canada's economic and social interests through the development and promotion of high-quality standards.

International Standards Organization (ISO): The International Standards Organization (ISO) is an independent, non-governmental international organization dedicated to developing and publishing a wide array of global standards. Established in 1947, ISO brings together experts from across the globe, encompassing diverse sectors such as industry, academia, and government. The primary mission of ISO is to foster innovation and provide solutions to global challenges by creating standards that ensure quality, safety, efficiency, and interoperability across various industries.

ISO's standards serve as a universal framework that facilitates seamless interaction and integration between different systems, technologies, and processes. By harmonizing technical specifications and guidelines, ISO ensures that products and services are reliable and compatible on a global scale. This is particularly crucial in today's interconnected world, where interoperability is key to fostering international trade, enhancing technological development, and addressing complex global issues.

With over 23,000 published standards covering everything from technology and manufacturing to healthcare and agriculture, ISO plays a pivotal role in setting the benchmark for best practices worldwide. Each standard is developed through a consensus-driven process, involving a multitude of stakeholders to ensure comprehensive and balanced guidelines. This collaborative approach not only enhances the relevance and applicability of the standards but also promotes their widespread adoption and implementation.

In the realm of smart cities, ISO has been instrumental in developing standards that address the need for coordinated efforts among municipalities. By establishing a unified framework, ISO enables cities to effectively collaborate, share knowledge, and implement smart city initiatives that are both scalable and sustainable. These standards ensure that smart city projects are not isolated endeavors but part of an interconnected network that maximizes their collective impact.

Individual Cities: Cities across Canada have been pursuing a variety of smart city initiatives tailored to their local needs and priorities. For example, the City of Toronto has implemented a smart traffic signal system that uses real-time data to optimize traffic flow and reduce congestion. They have also piloted smart waste management sensors to improve the efficiency of trash collection.

The City of Calgary has launched a smart parking system that allows drivers to locate and pay for parking spaces through a mobile app. This has helped reduce circling and idling, cutting greenhouse gas emissions.

The City of Montreal has deployed a network of environmental sensors around the city to monitor air quality, noise levels, and other metrics. This data is used to inform urban planning and policy decisions.

The City of Vancouver has invested in smart street lighting that automatically adjusts brightness based on pedestrian and vehicle activity. This has decreased energy usage and maintenance costs.

Other organizations: Municipalities and Regional Governments
Local governments across Canada play a significant role in smart city standardization. These municipalities often collaborate with federal and provincial governments, private sector partners, and international bodies to adopt and standardize smart city technologies.

6.2.4 Smart City Initiatives in Canada

Smart city initiatives in Canada encompass a wide range of topics, each aimed at improving urban living through the integration of technology, data, and innovative practices. Below are some key areas of focus:

- **Frameworks**

Frameworks provide the overall guiding principles and goals for smart city development. They establish the vision, objectives, and strategic directions for cities to follow. Canadian frameworks often emphasize sustainability, inclusivity, and resilience. They serve as blueprints for planning and implementing smart city projects, ensuring that initiatives align with broader policy goals and community needs.

- **Connectivity and Infrastructure**

Smart cities rely on robust digital infrastructure to support high-speed internet, IoT devices, and advanced communication networks. Initiatives in this area focus on expanding broadband access, implementing 5G technology, and enhancing public Wi-Fi availability. This connectivity is crucial for enabling other smart city applications, such as smart transportation and public safety systems.

- **Data and Analytics**

The collection, analysis, and use of data are at the core of smart city projects. Canadian cities are investing in platforms and tools that enable real-time data monitoring and decision-making. This includes open data portals that allow public access to government data, fostering transparency and community engagement. Advanced analytics help optimize city operations, from traffic management to energy consumption.

- **Mobility and Transportation**

Smart transportation initiatives aim to improve the efficiency, safety, and sustainability of urban mobility. This includes the development of intelligent transportation systems (ITS), autonomous vehicles, and shared mobility services. Cities like Toronto and Vancouver are piloting projects that use data analytics to reduce congestion, enhance public transit, and promote the use of electric vehicles and bike-sharing programs.

- **Energy and Environment**

Sustainability is a critical aspect of smart city initiatives. Canadian cities are adopting smart grids, renewable energy sources, and energy-efficient buildings to reduce carbon footprints. Environmental monitoring systems track air quality, water usage, and waste management, helping cities to manage resources more effectively and respond to environmental challenges.

- **Public Safety and Security**

Enhancing public safety through technology is a key priority. Smart surveillance systems, emergency response management, and predictive policing are some of the areas being developed. These technologies enable faster response times, better coordination among emergency services, and improved crime prevention strategies, ensuring safer urban environments.

- **Health and Well-being**

Smart city initiatives also focus on improving the health and well-being of residents. This includes telehealth services, health monitoring systems, and community wellness programs. Cities are leveraging technology to provide better healthcare services, promote active living, and address social determinants of health.

- **Citizen Engagement and Services**

Engaging citizens and improving public services through digital platforms is an essential component of smart city initiatives. E-government services, mobile apps, and social media channels are being used to enhance communication between citizens and government. Participatory platforms enable residents to contribute to decision-making processes, fostering a sense of community and collaboration.

- **Governance and Policy**

Effective governance and supportive policies are crucial for the success of smart city initiatives. Canadian cities are developing regulatory frameworks that address data privacy, cybersecurity, and ethical considerations. Collaborative governance models involve multiple stakeholders, including public, private, and academic sectors, ensuring that smart city projects are well-coordinated and inclusive.

6.3 Research Methodology: Case Studies

ISO is a non-profit standards organization with representatives from industry, government and non-profits. Their smart cities standards project involved subject matter experts from 15 Canadian municipalities. In 2014, a report by Infrastructure Canada found that lack of common standards was hindering collaboration on smart city solutions between levels of government and sectors.

This section dives into specific case studies, analyzing the work of a chosen standardization body.

6.3.1 Strategic Level

These standards provide guidance for city leadership on developing a comprehensive smart city strategy, including identifying priorities and establishing goals.

6.3.2 Process Level

This level focuses on procuring and managing smart city projects, ensuring efficient implementation and resource allocation.

6.3.3 Technical Level

Technical specifications are presented, outlining the specific requirements for implementing smart city products and services, fostering interoperability and functionality.

6.3.4 Smart City Calgary

The City of Calgary's Technology Integration Centre (TIC) is advancing as a centre of innovation, prepared to expand its influence globally. The vision is to foster a future where TIC's innovative spirit and collaborative approach impacts smart city initiatives worldwide. Here's the roadmap for this expansion:

- **Sharing Best Practices and Innovations**: TIC has developed exemplary projects serving as models for cities globally. Through conferences, publications, and online platforms, TIC aims to inspire and guide municipalities in their digital transformation journeys, creating a global network of smart cities.
- **Forging International Partnerships**: Central to TIC's expansion is establishing partnerships with international technology firms, academic institutions, and city governments. These alliances enable joint projects, resource sharing, and application of global expertise to address urban challenges and develop scalable solutions.
- **Participating in Global Innovation Networks**: Active engagement in global innovation networks and smart city alliances is crucial for TIC. These platforms promote collaboration, knowledge exchange, and joint innovation projects, ensuring TIC remains at the forefront of technological advancements.
- **Showcasing Success Stories Internationally**: By presenting achievements at international forums and tech summits, TIC amplifies its impact and attracts global interest, positioning Calgary as a leader in smart city innovation.
- **Offering Consultancy and Expertise**: Leveraging its proficiency in smart city solutions, TIC supports other cities in their digital transformation journeys. Consultancy services in smart infrastructure, data analytics, and public engagement bolster TIC's reputation and assist communities worldwide.
- **Leveraging Digital Platforms for Global Outreach**: Digital platforms play a pivotal role in amplifying TIC's insights and innovations globally. Webinars, online

courses, virtual tours, and social media engagement build a robust global community of innovators.
- **Scaling Successful Solutions Globally**: Many solutions developed at TIC are scalable for diverse urban settings. Collaboration with international partners facilitates global deployment, improving urban living standards worldwide.

The Technology Integration Centre is poised to transcend its local origins, significantly contributing to the global smart city movement. By sharing innovations, forging partnerships, and engaging in global networks, TIC will shape the future of cities worldwide, creating smarter, more connected, and resilient urban environments globally.

6.4 Conclusions

Patterns of Smart City Growth in Canada: Developing Smart, Sustainable Urban Centers

Canada, with its vast territory and diverse urban landscapes, presents a unique canvas for the evolution of smart cities. The concept of smart cities, integrating advanced technologies to enhance urban living and sustainability, is gaining momentum across the country.

In 2024, the Canadian government announced a landmark CAD$2 billion AI initiative, underscoring its commitment to integrating cutting-edge technologies into urban development strategies. This initiative not only accelerates the deployment of AI-driven solutions but also reinforces Canada's position as a global leader in smart city innovation.

Canada, divided into 10 provinces and three territories, boasts cities that are embracing innovative solutions in transportation, energy management, digital infrastructure, and citizen engagement. From Vancouver to Toronto, and from Calgary to Montreal, these cities are pioneering smart and sustainable urban centers that set global benchmarks for development.

For example, Calgary, for instance, has launched the 5G Discovery Zone[*]. This initiative aims to establish a 5G-enabled test environment in downtown Calgary, fostering innovation in connectivity and digital infrastructure. Such initiatives are crucial in positioning cities like Calgary at the forefront of smart city development in Canada.

The expansive geography of Canada necessitates tailored approaches to smart city development, focusing on local needs and opportunities. By leveraging technologies such as IoT (Internet of Things), AI (Artificial Intelligence), and data analytics, Canadian cities are paving the way for smarter, more connected communities. These technologies not only improve efficiency in resource utilization but also empower residents and businesses to actively participate in shaping their urban environments.

Moreover, Canada's commitment to sustainability aligns with the smart city agenda, promoting green infrastructure, renewable energy adoption, and low-carbon transportation options. By integrating these principles into urban planning and governance, cities are enhancing quality of life while fostering economic vitality and environmental stewardship.

As we look ahead, the journey towards smart cities in Canada involves collaboration among government, industry, academia, and citizens. Establishing robust standards and frameworks for data privacy, cybersecurity, and interoperability will be crucial in ensuring the reliability and scalability of smart city solutions.

In conclusion, Canada's expansive terrain and commitment to sustainability make it an ideal incubator for smart city technologies. By embracing innovation and collaboration, Canadian cities are poised to lead the way in building smart, sustainable urban centers that enhance livability, economic vitality, and environmental stewardship.

6.5 Revision Questions with Answers

What are the primary objectives of implementing smart city standards in Canada?

The primary objectives include enhancing efficiency, sustainability, and livability through the integration of technology and data-driven decision-making.

How does the ISO 37106 standard contribute to the development of smart cities in Canada?

This standard provides a set of standardized indicators for city services and quality of life. It is widely adopted by Canadian cities to benchmark performance. ISO 37106 provides guidelines for sustainable and resilient urban development, helping Canadian cities improve infrastructure, governance, and services to meet the needs of citizens effectively.

What role do standards such as ISO 37122 play in ensuring interoperability among smart city systems in Canada?

ISO 37122 establishes common metrics for evaluating smart city performance, facilitating interoperability among diverse systems and ensuring compatibility and data exchange across different urban areas in Canada.

How do smart city standards address concerns regarding data privacy and security in Canada?

Smart city standards incorporate provisions for robust data governance frameworks, encryption protocols, and privacy safeguards to protect citizen data and ensure compliance with Canadian privacy laws such as the Personal Information Protection and Electronic Documents Act (PIPEDA).

What are some challenges associated with the adoption of smart city standards in Canada, and how can they be addressed?
Challenges may include financial constraints, lack of standardized approaches, and community engagement. These challenges can be addressed through public–private partnerships, knowledge-sharing networks, and stakeholder involvement to foster innovation and collaboration in smart city initiatives across Canada.

Adopted International Smart Cities Standards

- **ISO 37120**: This standard focuses on indicators for city services and quality of life in sustainable cities and communities. Canada's participation underscores its commitment to leveraging global frameworks to improve urban sustainability, city planning, and quality of life for residents.
- **ISO 37122**: Focuses on sustainable cities and smart infrastructure, providing guidelines on the integration of ICTs in urban areas.
- **ISO 37123**: Addresses resilient cities, helping cities to prepare for and recover from various disruptions.

Acknowledgements 1. Rechie Valdez, Minister of Women and Gender Equality. Secretary of State (Small Business and Tourism). MP for Mississauga—Streetsville, Canada. 2. Chris Zhou, Director of Communications to the Minister of Small Business, Canada. 3. Jibril Hussein, Acting Director of Policy, Minister of Small Business, Canada. 4. Mia Ly, Content creator, Canada. 5. Dr Christina Yan Zhang, CEO of The Metaverse Institute. Co-chairman of The TaskGroup on Pre-standardisation for the CitiVerse, International Telecommunication Union. 6. Prof NK Goyal, Chairman Emeritus, TEMA-Telecom Equipment Manufacturers Association. President CSAI-Cyber Security Association of India. President, CMAI Association of India. Global Director of Cyber Security IHRO-International Human Rights Organisation. 7. Ernesto Faubel-Cubells, Chair of European Digital Infrastructure Consortium (EDIC) on Local Digital Twins towards the CitiVERSE. Head of Smart City & Data Management Department. Valencia, Spain. 8. Brenda Bailey, BC Finance Minister. Ex-Minister of Jobs, Economic Dev and Innovation (JEDI), Canada.

References

International Organization for Standardization (2017) ISO 37106: Sustainable cities and communities—guidance on establishing smart city operating models for sustainable communities. ISO

International Organization for Standardization (n.d.) ISO 37120: Sustainable cities and communities—indicators for city services and quality of life. Retrieved from https://www.iso.org/standard/62436.html

Standards Council of Canada (2018) Smart cities standards roadmap. Standards Council of Canada. https://www.scc.ca/en/news-events/news/2018/smart-cities-standards-roadmap

Val Wise/Valeriy Natrus/ is a leading Marketing and Innovation Consultant, recognized as a UN U4SSC Expert and a Top Innovation Voice in Smart Cities. With a sharp focus on strategy, generative AI, and digital transformation, he advises cities and companies across Canada and globally. Val is also the insightful host of a popular podcast and a highly sought-after international speaker. His groundbreaking work seamlessly integrates cutting-edge AI tools with smart city solutions, consistently driving innovation across both public and private sectors.

7. The Spanish Vision of Smart Cities and Smart Tourist Destinations Through Standards

Ramón Ferri Tormo, Beatriz de Esteban Martín, and Tania Marcos Paramio

Abstract

In recent years, Spain has played a prominent role in the standardization and establishment of standards in the field of smart cities, particularly in the area of smart tourism. Standardisation was one of the strategic pillars of the National Smart Cities Plan, and the Spanish Association for Standardization (UNE) established in 2012 the Standards Committee CTN 178 'Smart cities', initially promoted by the Secretary of State for Telecommunications and the Information Society. The Ministry of Tourism, through SEGITTUR, promoted the creation of the Subcommittee 5 'Smart Destinations' (SC5) within it with the aim of creating a consistent framework for smart tourism destinations. Through the Standards Technical Committee CTN-UNE 178 and the set of standards from Subcommittee 5 'Tourist Destinations', Spain has led the creation of standards related to smart tourism destinations. Additionally, it has closely collaborated with other standardization entities such as the International Telecommunication Union (ITU) and the International Organization for Standardization (ISO). Some of these standards, such as UNE 178104 on interoperability of smart city platforms or UNE 178503 on semantics applied to tourism, have gained broad international following and have led to further works in the Initiative of United Nations 'United for Smart Sustainable

R. F. Tormo (✉)
Valencia, Spain

B. de Esteban Martín
SEGITTUR, Madrid, Spain
e-mail: beatriz.deesteban@segittur.es

T. M. Paramio
Asociación Española de Normalización (UNE), Madrid, Spain
e-mail: tmarcos@une.org

Cities' (U4SSC) related to UN Sustainable Development Goals. This chapter analyzes the role of Spain in the standardization of smart city destinations, highlighting the reasons for their creation and providing an overview of the structure, as well as the advantages of these standards and their impact in the global context.

Keywords

Smart tourism destination · Standards · Digital transformation · Sustainability · Innovation

7.1 Introduction

The tourism industry plays a fundamental role in the global economy and in promoting cultural diversity and intercultural understanding. However, the growth and evolution of this sector have generated challenges related to service quality, environmental protection, and preservation of cultural heritage. To address these issues and promote sustainable tourism development, standards have been created worldwide.

In the context of the digital revolution, smart cities have become a key priority to drive improvements in the quality of life for residents, foster sustainable practices, and promote efficient urban management. The adoption of advanced technologies and the integration of innovative solutions enable harnessing the potential of digitization to address urban challenges and enhance the citizens' experience. The goal is to create smarter, connected, and sustainable urban environments where technology is strategically used to optimize public services, mobility, resource utilization, and citizen participation. Through the use of sensors, connected devices, and information systems, smart cities can collect real-time data, analyze it, and make informed decisions to provide greater efficiency, safety, and quality of life to their residents. By fostering interaction and collaboration among different city stakeholders, the aim is to promote innovation, creativity, and citizen participation in decision-making and the construction of a more sustainable and livable urban future. In this context, Spain has played a significant role in the standardization of smart tourism destinations.

Spain, through the Standards Committee CTN-UNE 178 of the Spanish Association for Standardization (UNE), has played a prominent role in the standardization of smart cities and smart tourism destinations. Spain's commitment to the creation of technical standards has led the way in defining requirements and best practices for the development of these destinations. In collaboration with other standardization entities such as the International Telecommunication Union (ITU) and the International Organization for Standardization (ISO), efforts have been made to harmonize and adopt standards globally.

These Spanish standards have gained broad international following, contributing to the advancement and improvement of the tourism sector globally. The adoption of these standards provides a common framework and shared language for project development and implementation of solutions in tourist destinations. Moreover, they promote interoperability of systems, fostering collaboration among different stakeholders and facilitating service integration for the benefit of citizens and tourists.

In conclusion, Spain, through CTN-UNE 178 and its Subcommittee 5 on Tourist Destinations, has led technical standardization in the field of smart tourism destinations. By creating technical standards, requirements and best practices have been established and adopted at both national and international levels. These standards have contributed to improving efficiency, sustainability, and the quality of services in tourist destinations, positioning Spain as a reference in the development of smart cities and tourist destinations.

7.2 Background

The promotion of standardization in the field of smart cities has become a global priority. In this regard, Spain has actively contributed to the development of standards at both national and international levels. The Standards Committee CTN-UNE 178, of the Spanish standardization entity UNE, has been responsible for leading standardization efforts in Spain. Additionally, it has collaborated with international organizations such as ITU and ISO to promote harmonization and the adoption of standards at a global level. To share the Spanish experience and lessons learned, CTN-UNE 178 committee members participate actively at the UN initiative United for Smart Sustainable Cities (U4SCC), leading the thematic Group on City Platforms and the works on Smart Tourism Destinations.

7.2.1 The Importance of Standardization

Quality, safety, and sustainability within tourist destinations are all dependent on standards in the tourism industry. Positive experiences for tourists are ensured by these standards, which provide a structured framework with essential requirements that destinations must adhere to.

Standardization was a fundamental component of the National Smart Cities Plan, which included buildings, stations, ports, airports, islands, rural areas and tourist destinations. In 2011, the Spanish Association for Standardization (UNE) established the Standards Committee CTN 178 'Smart Cities,' initially championed by the Secretary of State for Telecommunications and the Information Society, aligning with Spain's Digital Agenda. Furthermore, the Ministry of Tourism, in collaboration with SEGITTUR, spearheaded the establishment of Subcommittee 5 'Smart Destinations' (SC5) with the aim of establishing a cohesive framework for innovative tourism destinations.

Subcommittee 5 is tasked with defining specific guidelines and requirements for tourist spots in order to guarantee pleasant encounters for those who visit. Tourism infrastructure, service offerings, environmental stewardship, and cultural heritage preservation are some of the things these standards encompass. Quality benchmarks and exceptional services can be achieved by adhering to these standards (UNE 2021).

Enhancing the quality of tourist destinations yields positive outcomes for both visitors and local stakeholders engaged in the tourism sector. High-end amenities boost client contentment, causing them to pass along the spot to their buddies. Furthermore, locations that adhere to established guidelines stand a good chance of attracting a greater number of visitors and establishing themselves as trustworthy and trustworthy destinations (SEGITTUR 2015).

The concept of sustainability requires a delicate balance between economic advancement, environmental conservation, and safeguarding cultural heritage. Responsible natural resource management, reduction of environmental footprints, community engagement, and cultural heritage preservation are some of the sustainable practices fostered through the establishment of requirements related to sustainable practices in tourist destinations.

Tourism destinations can safeguard natural and cultural resources by embracing these standards. Furthermore, a focus on environmental stewardship enables destinations to stand out and appeal to a progressively mindful group of tourists who value ethical travel habits.

Another rationale for the formulation of standards by Subcommittee 5 is to align with international standardization efforts and garner recognition on a global scale. UNE's CTN 178 guides the development of these standards, encouraging collaboration with relevant global tourism bodies. Having Spanish standards in line with global standards encourages collaboration between tourist spots around the globe.

Common standards foster coherence and compatibility among destinations, facilitating comparability and the exchange of best practices. They also facilitate comparability and the exchange of best practices. The compliance with these standards positions destinations as esteemed and sustainable destinations on the international stage. These standards are effective in addressing the evolving needs of the tourism industry and promoting sustainable practices for the benefit of both tourists and destination stakeholders because of the quality of research underpinning them.

7.2.2 Importance and Utility of Tourist Destination Standards

Standardization plays a pivotal role in addressing the primary challenges faced by the tourism sector. By implementing and adhering to comprehensive standards, tourist destinations can ensure sustainable development, enhance competitiveness, and provide high-quality experiences for visitors. The following outlines the critical aspects where standards are indispensable (UNE "Apoyo de la normalización al sector turístico" https://www.une.org/salainformaciondocumentos/Informe%20UNE%20Normalizaci%C3%B3n%20tur%C3%ADstica.pdf):

- Enhancing competitiveness: Standards help destinations stand out by ensuring quality and sustainability, attracting more tourist, and driving regional economic and social development.
- Preserving sustainability: Aligning with SDGs, standards promote sustainable practices, resource management, and conservation, protecting natural and cultural heritage for future generations.
- Digitalization and intelligence: Standards support digital transformation and smart management, enhancing operational efficiency and visitor experiences, crucial for global competitiveness.
- Differentiation through Quality: High-quality standards guarantee reliable and authentic tourism experiences, leading to higher satisfaction, positive recommendations, and new visitors.
- Diversification of offerings: Standards encourage diversifying tourism products, attracting a broader audience and supporting local economies through varied opportunities.
- Increasing Tourist Spend: Focusing on high-quality, professional services increases tourist expenditure, boosting local economic prosperity.
- Protecting Authenticity: Standards protect cultural an natural heritage from overexploiting, offering genuine experiences that appeal to culturally conscious travelers.
- Informed Decision-Making: Standards provide a framework for destination managers to make informed decisions, optimize resources, and achieve tourism development goals effectively.

The implementation of standards plays a critical role in multiple facets of tourism development. Now, let's delve deeper into how these standards specifically enhance the competitiveness of tourist destinations in the global market.

Enhancing the competitiveness of tourist destinations in the global market: By fulfilling the established requirements and guidelines, destinations can differentiate themselves and demonstrate their commitment to quality and sustainability. This contributes to attracting a greater number of tourists and generating a positive image, which in turn drives economic and social development in the region.

Tourism can exert pressure on the natural and cultural resources of a destination. The standards developed by this committee promote the protection and preservation of cultural and natural heritage, avoiding overexploitation and irreversible deterioration. These standards foster responsible resource management, the promotion of sustainable practices, and respect for local identity and traditions.

Moreover, these standards provide a guarantee of quality and trust for tourists. When a destination complies with the established standards, visitors can trust the quality of services, safety, and authenticity of the tourism experience. This translates into higher tourist satisfaction and generates positive recommendations, benefiting the reputation and attraction of new visitors.

These standards provide a reference framework that helps destination managers make informed decisions, optimize available resources, and effectively achieve tourism development objectives.

Spain's leadership in this area has been recognized by the United for Smart Sustainable Cities (U4SSC), a global UN initiative coordinated by ITU, UNECE and UN-Habitat, and supported by other 16 UN agencies, that help support the development of institutional policies and strategies which encourage the use of digital technologies to facilitate digital transformation and ease the transition to smart sustainable cities. Spain is leading the work on City platforms for the development of several U4SSC deliverables, including the one on 'Smart tourism: A path to more secure and resilient destinations' and starting to work on Smart Destination Platforms. It has also been relevant to participate ITU's webinars on digital transformation for cities and communities, highlighting the *Episode #19: Tourism in smart cities: Reimagining the road to digital tourism*, organized jointly by ITU, UNWTO and UNE.

7.3 Research Methodology: Subcommittee 5 'Tourist Destinations Standards

The Standards Technical Committee CTN 178 was established in 2012 with initial support from the then Ministry of State for the Information Society and the Digital Agenda. At that time, standardization was one of the pillars of the National Plan for Smart Cities, which later transitioned into the National Plan for Smart Territories.

The Standards Technical Committee CTN-UNE 178 on Smart Cities is a public–private collaboration group of experts on smart sustainable cities responsible for laying the foundations for addressing the challenge of digital transformation in smart cities and tourist destinations, including smart communities, rural areas, and island territories. CTN 178 involves more than 700 experts from 237 entities, representing all the stakeholders involved in the challenge of transforming cities into smart cities and intelligent tourist destinations. The goal is to strengthen the digitization and sustainability of key verticals of services provided by the city while adopting a highly cross-cutting and holistic perspective of the different processes and agents that need to be involved in the complex city ecosystem.

In the realm of smart cities, Spain has taken a pioneering role in establishing standards for open data and the effective management of municipal assets, infrastructure, and performance metrics across various public service domains, including water, waste, transportation, telecommunications, and energy. This extends to areas such as street lighting, irrigation, and ensuring universal accessibility. Our advanced expertise in achieving interoperability among smart city platforms, intelligent buildings, and internal urban infrastructure components like transit stations, ports, and airports, has been shared with the

International Telecommunication Union (ITU). This collaboration has resulted in the successful development of ITU Recommendations, further facilitating the global expansion of Spanish companies.

Furthermore, Spain emerged as a leading participant in the Sectoral Forum dedicated to Sustainable and Smart Cities and Communities established by the three European standardization bodies, namely European Committee for Standardization (CEN), European Committee for Standardization (CENELEC), and European Telecommunications Standards Institute (ETSI), of which UNE is a member, where served as the representative for this forum in its interactions with the European Commission's European Innovation Partnership on Smart Cities and Communities (EIP-SSC).

Spanish Standardization also contributes to relevant international and European initiatives such as U4SSC (with ITU, UNECE, UN Habitat and other United Nations agencies) and Living-in EU (promoted by European Commission, European Committee of the Regions, Eurocities, Open and Agile Smart Cities (OASC) and others).

The Spanish Standards Committee CTN-UNE 178 is organized into 7 Subcommittees, each led by different public administrations, focusing on standardization in specific thematic areas. These areas include infrastructure (SC1), indicators and semantics (SC2), mobility and transport platforms (SC3), sustainability (SC4), tourist destinations (SC5), land-use planning and public services (SC 6) and datification and public services (SC7).

More than 30 Spanish standards developed within CTN-UNE 178 focus on enhancing services for citizens and visitors while promoting dialogue between government bodies and industry, serving as a reference for public procurement. It is worth highlighting that CTN-UNE 178 enjoys the support of the Spanish Federation of Municipalities and Provinces (FEMP), the Network of Smart Cities (RECI), the Network of Smart Tourist Destinations (Red DTI), the Network of Urban Initiatives (RIU), and the Network of Science and Innovation Cities (Red Innpulso).

The standards developed by these Subcommittees cover a wide range of areas within tourism. They range from the management of smart cities (UNE 178104:2017) and requirements for smart buildings to be considered IoT nodes (UNE 178108:2017) to smart station and connection with the smart city platform (UNE 178109:2018). Other UNE standards related to smart cities address open data, territories and semantics applied to data.

Within Subcommittee 5 Tourist Destinations, standards have been developed for topics such as Wi-Fi for beaches, SEO/SEM or tourist Wi-Fi, data collection, exploitation and analysis model for tourism, among others (Fig. 7.1).

The Spanish standards that have emerged from CTN-UNE 178/Subcommittee 5 on Tourist Destinations are:

UNE 178501:2018 Smart Tourist Destination Management System. Requirements: This standard establishes requirements for implementing, maintaining, and enhancing a management system for smart tourist destinations, encompassing the criteria of governance, innovation, technology, universal accessibility, and sustainability. By setting clear criteria,

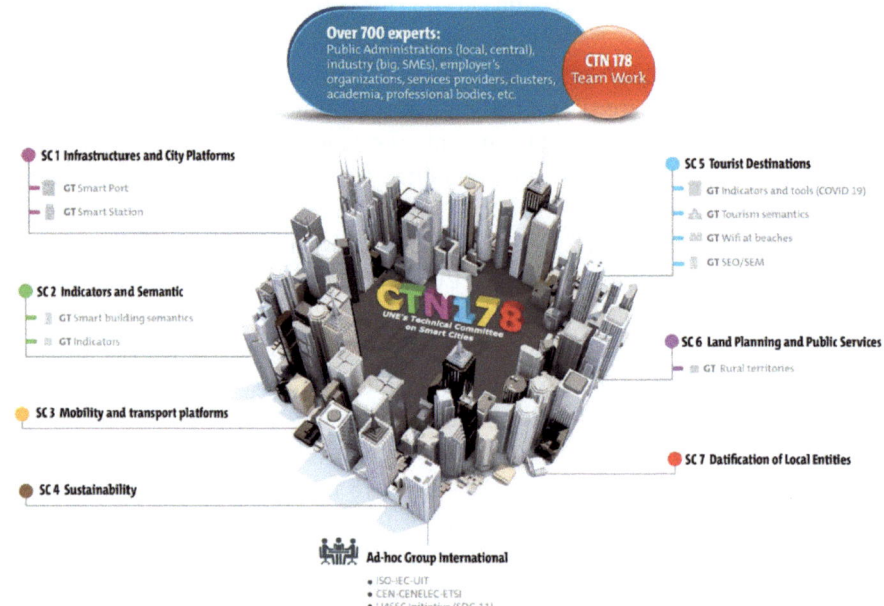

Fig. 7.1 Representation of the structure of the Spanish standards committee CTN-UNE 178 'Smart Cities'

it provides a framework for enhancing the management and quality of tourist services in smart destinations. Aligned with existing literature emphasizing effective management in sustainable tourism, the UNE 178501:2018 standard highlights the necessity of integrating governance, innovation, technology, universal accessibility, and sustainability criteria to ensure excellence in smart tourist destination management. Research design plays a crucial role in assessing the effects of implementing these requirements on operational efficiency, visitor satisfaction, and destination sustainability (Dogra and Kale 2020). Studies applying similar standards have demonstrated improvements in destination management, technology integration, and tourist experiences in smart destinations. The adoption of standardized requirements not only fosters excellence in tourist management but also stimulates innovation, efficiency, and competitiveness in smart tourist destinations, thereby contributing to the sustainable development of the sector.

UNE 178502:2022 Indicators and Tools for Smart Tourist Destinations: This standard determines and develops common objectives according to a defined continuous improvement strategy and makes the necessary changes to achieve these objectives. It allows for the measurement of progress and its comparison based on common criteria, while also establishing a set of tools and indicators to provide information on goals and parameters linked to activities or processes. Aligned with existing literature emphasizing the importance of measuring progress and setting common goals in the management of sustainable tourist destinations, the UNE 178502:2022 standard underscores the necessity of tools and metrics to assess the performance, effectiveness, and sustainability of tourist destinations. Research design plays a crucial role in evaluating the effectiveness of the established indicators and tools for measuring performance and effectiveness of implemented actions (Ivars-Baidal et al. 2021b). Studies applying similar metrics have demonstrated improvements in destination management, data-driven decision-making, and optimization of tourist experiences in smart destinations (Ivars-Baidal et al. 2021a). The adoption of standardized indicators and tools not only enables progress assessment and result comparison but also drives innovation, efficiency, and competitiveness in smart tourist destinations, contributing to the sustainable development of the sector.

UNE 178503:2022 Semantics Applied to Smart Tourist Destinations:

This standard holds significant importance in shaping the tourism landscape as it allows for the representation of pertinent destination information in a structured manner. By guaranteeing compatibility among different tourism platforms, such as those operated by cities, territories, and external entities, it enables smooth data sharing and integration. This compatibility enables all participants in the tourism industry to make the most of current technology, guaranteeing that tourists have access to the most relevant and current information about the destination (Angele et al. 2020).

Additionally, the semantic structure offered by this standard gives meaning to tourism activities, enabling machines to interpret them. This semantic enhancement not only improves the way humans and machines communicate but also enables more effective communication between machines themselves. As a result, travelers' experiences are enhanced as they can receive personalized suggestions, customized information, and effortless interactions with different digital services while on their trips. Additionally, the standard encourages innovation and cooperation within the tourism sector by establishing a common understanding of destination information among all stakeholders involved. Overall, the standard plays a crucial role in improving tourist experiences and streamlining destination management practices in the digital era.

UNE 178504:2022 Digital, Smart, and Connected Hotel (HDIC) to Smart Tourist Destination/Smart City Platforms. Requirements and Recommendations. This standard outlines requirements for transforming accommodation into digitally connected hotels within smart tourist destinations or cities, stressing bidirectional communication to enhance the tourist experience and ecosystem competitiveness, especially for hotels. Aligned with research emphasizing technology integration, it underscores digital transformation's role in improving guest experiences and hotel competitiveness. Effective communication between destinations and accommodations is crucial for smooth guest experiences and industry innovation (Buhalis and Leung 2018).

The standard aims to enhance tourist experiences and ecosystem competitiveness by providing guidelines for hotel transformation. Studies show positive changes in guest interaction, personalized services, and revenue. Standardized requirements drive innovation, efficiency, and guest loyalty, crucial for the success of smart tourism initiatives.

UNE 178505:2022 Framework for the Creation of Tourist Destination Websites and UNE 178506:2022 Methodology for Search Engine Optimization (SEO) of Tourist Destination Websites: The standards offer a comprehensive approach for creating and designing promotional websites for destinations, supporting developers in both development and maintenance tasks. Additionally, they define best practices for search engine optimization to improve crawling, indexing, positioning, user experience, and organic traffic performance on search engines (Shrestha et al. 2021). Research underscores the vital role of digital marketing in destination promotion, highlighting the significance of well-designed websites and effective SEO strategies. Implementing these standards enhances website usability, search engine rankings, and user engagement, ultimately increasing online visibility and attracting visitors. By providing structured guidelines, the standards streamline website development, optimizing online performance and user experience to drive successful destination marketing strategies.

UNE 178507:2022 Tourist Destinations. Applications of Wi-Fi Connection on Beaches: This standard outline general guidelines for Wi-Fi services in beach settings, aligning with research emphasizing the significance of connectivity in enhancing destination management. Studies confirm that Wi-Fi is crucial for improving visitor experiences and enabling data-driven decision-making (Almeida 2018). Dependable Wi-Fi enhances visitor satisfaction, operational efficiency, and destination competitiveness. Standardized guidelines ensure the highest service quality, facilitating effective data collection and informed decision-making for destination management. Research indicates that compliance with these standards enhances data accessibility, visitor involvement, and operational effectiveness, leading to more informed decision-making in beach destinations. These guidelines allow for data collection to improve decision-making by destination management entities. UNE 178508:2022 Model of Tourist Destination Applications (Apps) for Mobile Devices: This standard provides comprehensive guidance for developing mobile applications tailored to tourist destinations, with a focus on user-centered experiences crucial in the tourism industry. Emphasizing the significance of well-designed apps, research underscores the necessity for standardized guidelines to ensure intuitive, secure, and up-to-date information, thereby enhancing user satisfaction. By adhering to these guidelines, destinations can create apps that offer accurate information about destinations, activities, services, and events, ultimately improving the overall travel experience and fostering healthy competition. UNE 178509:2023 Collection, Exploitation, and Analysis Model of Data: This standard provides a comprehensive framework for data management in the tourism sector. This standard addresses data collection, processing, analysis, and utilization, which is crucial for making informed decisions, identifying trends, and enhancing decision-making in the strategic development of tourist destinations.

Various reports highlight how advanced data frameworks improve tourist satisfaction and operational efficiency (Abraham et al. 2019). These frameworks facilitate the integration of multiple data sources, ensuring comprehensive insights that support strategic tourism development. Implementing this standard not only enhances data management practices but also drives economic growth and innovation within the tourism sector.

Studies demonstrate that destinations with robust data management frameworks experience increased tourist arrivals and higher revenue generation. Furthermore, these practices foster the development of new technologies and services, enriching the overall tourist experience. By incorporating these improvements, your document will present a comprehensive literature review and a well-structured research design, significantly enhancing the overall quality and impact of your work..

UNE 178510 Intelligent Tourism Enterprise:

This standard specifies the requirements and recommendations for an Intelligent Tourism Company (ITE) to effectively address management, innovation, technology use, universal accessibility, and sustainability.

Implementing these standards is of great strategic importance, as it underscores the quality, innovation, and sustainability of smart tourist destinations, positioning Spain as a leader in this field. These standards are crucial for modernizing tourism infrastructure and aligning with global best practices. Research from the World Tourism Organization and various academic sources highlights the positive impact of these standards on tourist satisfaction and destination competitiveness.

Compliance with these standards not only enhances the tourist experience but also contributes to economic growth and technological advancement. Studies show that adopting smart city/destination standards and technologies can increase tourist arrivals and higher revenue generation for local economies, reinforcing the case for evidence-base digital transformation (Ji et al. 2024). Furthermore, these standards encourage innovation within the tourism sector, fostering the development of new technologies and services that improve the experiences of both visitors and residents.

UNE 178511:2023 Guide for the Application of the Layer Model of the Intelligent Destination Platform:

This standard provides essential guidance for the application of the requirements of the layer model for smart destination platforms, as specified in the Spanish Standards UNE 178104 and UNE 178502. This standard is essential to improve the integration and efficiency of the technological layers within smart destinations, thus enhancing data management, service provision and the tourist experience in general.

The importance of structured layer models in smart tourism is highlighted. Studies carried out by the World Tourism Organization and academic researchers highlight the importance of these models to guarantee perfect integration and interoperability of technologies in the field of tourism management. The importance of modularity and adaptability of layered models is highlighted, allowing destinations to keep up with technological advances and changing tourism needs.

The adoption of the UNE 178511:2023 standard not only increases the quality of tourist services, but also drives economic growth and technological innovation.

7.4 Other Standards Produced by CTN-UNE 178 'Smart Cities'

7.4.1 Subcommittee SC1—Infrastructures and Smart City/ Destination Platforms

UNE 178101:2015 Smart cities. Infrastructures. Public Service Networks. The UNE 178101 series of standards defines metrics applicable to the public services networks, in order to provide the best services to the citizens whilst providing maximum efficiency and easy integration within the environment, in the Smart and sustainable Cities framework. It has several parts defining metrics applicable to: Part 1—drinking water and wastewater public service networks of a city; Part 2—Waste Networks; Part 3—Transport Networks; Part 4—Telecommunication Networks, and Part 5—Energy Networks.

UNE 178102:2015 Smart cities. Infrastructures. Telecommunication systems. Part 1 of the UNE 178102 Standard series defines a multiservice city telecommunication network which supports the city infrastructures and services provided by the city council, as well as its composition and its classification in telecommunication systems and transmission networks. Part 3: Unified Communications Systems.

UNE 178107:2015 IN Guidelines on smart cities infrastructures. Access and transport networks. The series of Technical Reports, UNE 178107 IN on the support networks of a multiservice city network, complement UNE 178102 Standards series on: Fibre optics networks, Wireless Metropolitan Area Network (WMAN), Wireless Local Area Networks (WLAN), Wireless Sensor Networks (WSN), Safety, Security and Emergency mobile networks (SSE) and Radiolinks.

UNE 178104:2017 Comprehensive systems for a smart city management. Requirements of interoperability for a Smart City Platform. Spain was a pioneer in developing standardisation to guarantee interoperability in smart city platforms, through a layered model that will require open and standardised interfaces.

This Spanish standard was the root for the development of ITU-T Recommendations Y.4200: *Requirements for the interoperability of smart city platforms* and Y.4201: *High-level requirements and reference framework of smart city platforms*. In consequence, within the United Nations U4SSC initiative, Spain leads the Thematic Group on City Platforms, addressing IT solutions, smart energy management for health, interoperability of city architecture and digital and sustainable tourism, where a number of free deliverables have been produced.

More recently, precisely because of the importance of interoperability, Spain and Denmark have promoted, with the support of the European Commission (DG CNECT), the development of the forthcoming ITU Recommendation on Minimal Interoperability Mechanisms, based on the OASC MIMs.

UNE 178108:2017 Smart cities. Requirements for the application of UNE 178104 to smart buildings. Spain's national smart city plan identified buildings early on as an internal city object, which could provide relevant data to improve city governance and efficiency. This standard specifies the requirements for the integration of the information of an intelligent building, as an IoT node, with the city platform, according to the requirements of the standard UNE 178104. It inspired the development of ITU-T Recommendation L1370: *Sustainable and intelligent building services.*

UNE 178109:2018 Smart Cities. Smart Station and Connection to the Smart City Platform
The document aims to guide and facilitate the transformation of passenger railway stations into smart stations, increasing their level of usability, improving their management and creating value-added services for citizens.

UNE 178110:2024 Smart Ports. Requirements and Recommendations
The importance of ports as links in logistic and transport chains is indisputable in Spain, which has the longest coastline in the European Union (8.000 km). Under development, this standard will define the criteria and characteristics necessary to consider a port as "smart". It aims to guide and facilitate the transformation and evolution of ports towards the concept of Intelligent Ports, improving their management, their competitive level and creating added value services for the companies that develop their activities on or around the port, for the entire Logistics-Port Community, the areas of influence of the port and the city. It states the requirements of the smart port platform, in accordance with UNE 178104 and the layer model set out in UNE 178511.

UNE 178105:2017 Smart Cities. Infrastructures. Universal Accessibility in the Smart Cities
Spain's long history of standardisation in universal accessibility and design for all is reflected in the demand for the development of smart cities that are accessible and inclusive for all citizens and visitors. This standard provides guidance for inclusion of accessibility and design for all criteria in smart cities, according to Spanish legislation. This standard aims to contribute to and establish a framework for developing standardised indicators/criteria to assess whether and to what extent a smart city or community is accessible. These indicators are suitable for application in very different cities and at different times, determining and identifying those elements and areas in which it must be present, establishing a general framework. They must consider both the attributes of the smart city or community itself, i.e., each of the needs it must meet to be "smart", and its specific requirements (equality and social inclusion, quality of life, environmental sustainability, productivity and Information and Communications Technologies).

7.4.2 Subcommittee SC2—Indicators and Semantics

UNE 178201:2016 Smart cities. Definition, attributes and requirements. This standard proposes a formal definition of the "Smart City" concept, identify the attributes that characterise it, the necessary requirements for a city to be considered a Smart City and finally, to describe a semantics of City that allows a coherent definition of standardised, consistent and comparable indicators over time and between cities. These indicators should provide a uniform approach to what is measured and how this measurement is to be carried out. It is useful not only for the definition of the concept itself, but also as a basis for common semantics and a connecting element between the ICT infrastructures, metrics and policies of Smart Cities.

UNE 178202:2016 Smart cities. Management indicators based on balanced scorecard. This standard defines and establishes a set of management indicators for the creation of a management scorecard to guide and measure the performance of urban services and the quality of life in the city.

UNE 178204:2021 Smart cities. Semantics applicable to data and information from building monitoring and their integration into higher-level living units. This standard defines the semantic bases necessary to exchange data and information between the elements of the building, with other buildings or higher coexistence units such as the city, in accordance with the UNE 178108 Standard and the International Recommendation ITU-T L.1370, in which an IoT node is identified as its management element. However, the semantic bases covered by this standard could be used by any other building data management element.

This Spanish standard has encouraged SEGITTUR to lead the development of future ISO 20525 "Tourism and related services. Semantics applied to tourism destinations" within the Technical Committee ISO/TC 228 "Tourism and related services", which is currently under development. https://www.iso.org/committee/375396.html

7.4.3 Subcommittee SC 4—Sustainability (Formerly Energy and Environment)

UNE 178401:2017 Smart cities. Street lighting. Degrees of functionality, zoning and control arquitecture. This standard establishes the requirements for actuation, dimming and remote management systems for outdoor lighting installations, as well as to determine their degrees of functionality in the smart city. Special consideration is given to energy savings with high efficiency luminaires and automatic lighting control, with adjustments based on, for example, daylight incidence, road type or occupancy.

UNE 178402:2015 Smart cities. Management of basic services and water and electricity supply in smart ports. This standard proposes the implementation of a remote management system to improve the management of port energy supplies and services to minimise consumption and the impact on the environment of the port and the city, and also to be able to interact with the city in which the port is located. This Spanish standard was the root for the development of ITU-T Recommendation Y.4209 'Requirements for interoperation of the smart port with the smart city'.

UNE 178405:2018 Smart Cities. Environmental sensoring. Smart irrigation system. Water demand is expected to increase due to population growth in cities, which also need to provide more green and landscaped areas or the well-being of their citizens. This standard helps to optimise the irrigation of such areas, providing savings in water consumption and also in the associated electricity.

7.4.4 Subcommittee SC 6 'Land Use and Public Services'

UNE 178301:2015 Smart Cities. Open Data. Open data and the re-use of public sector information are two essential aspects on the road to smart cities and open government. This standard facilitates of open data projects and improves their management, providing metrics and indicators and establishing a list of priority datasets and vocabularies. So, open data complies with accessibility requirements, can be reusable and exploitable by stakeholders, boosting innovation in the territories thanks to the generation of new products and services.

UNE 178303:2015 Smart Cities. City asset management. Specifications. The system described allows local authorities, according to their strategic plan, to optimally, sustainably and efficiently manage their assets, their performance, risks and associated costs throughout the entire life cycle of the different assets. Among other advantages, this document allows the local government to obtain accurate information on the state of conservation and the useful life of the city's elements and facilities, optimising acquisition and maintenance costs. It also facilitates adequate medium and long-term planning of new acquisitions and replacements of assets. In the eyes of citizens, it is a tool for greater transparency in local government management.

UNE 178601:2022 Smart Territories. Definition, Attributes and Requirements. This standard defines the political-administrative level to be built, using existing structures in Spain for the specificity of the Intelligent Territories, around small rural municipalities. It applies to any municipality with a smaller population (≤ 5.000 hab.) that undertakes to measure its performance in a comparable and verifiable way and to the political-administrative institutions in which they can be grouped to form Smart Territories, as well as contributing to the development of applications of institutions that allow a better understanding of that territory and its smaller population municipalities.

7.4.4.1 Participation and Experts Engagement

Much of the success of the standardization activity of Committee CTN-UNE 178 is due to the public–private collaboration, as more than 700 Spanish experts from different public and private sectors and with very different professional and societal profiles, contribute their experience.

The great involvement in the committee of city leaders and tourist destinations managers, together with sectoral associations and National networks, endorses the usefulness of the standardization work carried out. Also, participation of Spanish experts in the International and European standardization work serves to share experiences and to keep national standards up to date and enrich them. In this sense, it is also important to highlight the Spanish contribution to the Living-in.EU initiative in Digital Twins and Minimal Interoperability Mechanisms.

"In short, in Spain there is a common strategy and an aligned demand and supply. The difference is not only what is done, but how it is done, and Spain has a model to export; we have everything we need to be a success story" [Enrique Martínez Marín, Chairman of CTN-UNE 178].

7.5 Conclusions

Spain has been a key player in the development of smart cities and smart tourism destinations by taking the lead in the Standards Technical Committee CTN-UNE 178 "Smart Cities" and Subcommittee 5 on Tourist Destinations. This project has created a comprehensive framework that enables the efficient execution of smart tourism destinations throughout the nation. In creating technical guidelines, Spain has emphasized excellence, creativity, and environmental responsibility in destination management, thereby improving its global competitiveness in the tourism sector.

Spain has worked together with international standardization organizations like the International Telecommunication Union (ITU) and the International Organization for Standardization (ISO) to align and encourage the worldwide acceptance of standards. These initiatives have resulted in the establishment of globally accepted standards that improve effectiveness, sustainability, and overall quality of services in tourist destinations.

In summary, Spain, through CTN-UNE 178 and its Subcommittee 5 on Tourist Destinations, is at the forefront of technical standardization initiatives for smart tourism destinations. Spain is a leader in the creation of smart cities and tourist destinations, emphasizing the importance of systems that can work together and cooperation among all involved parties for the benefit of both residents and visitors.

7.6 Revision Questions with Answers

1. Why did Spain establish Subcommittee 5 'Smart Destinations' within the Technical Committee for Standardization CTN-UNE 178 'Smart Cities'?

    Answer: Spain established Subcommittee 5 'Smart Destinations' to create a consistent framework for smart tourism destinations and to promote quality, innovation, and sustainability in destination management.

2. How has Spain collaborated with international standardization entities in the development of smart tourism destination standards?

    Answer: Spain has collaborated with international standardization entities such as the International Telecommunication Union (ITU) and the International Organization for Standardization (ISO) to harmonize and adopt smart tourism destination standards globally.

3. What is the significance of UNE 178503, the standard on semantics applied to tourism, in the context of smart tourism destinations?

    Answer: UNE 178503 facilitates interoperability of tourism platforms and enhances the efficiency of information exchange in smart tourism destinations by providing a semantic structure to tourism activities.

4. How do the standards developed by Subcommittee 5 contribute to the competitiveness of tourist destinations in the global market?

    Answer: These standards enhance the competitiveness of tourist destinations by promoting quality, sustainability, and trust. Compliance with these standards helps destinations differentiate themselves and attract more tourists, driving economic and social development.

5. How do standards enhance the competitiveness of tourist destination in the global market?

    Answer: Standards make tourist spots more competitive in the global marketplace by making sure they meet established standards and guidelines, which makes them stand out and shows their dedication to quality and sustainability. This attracts a higher number of tourists, generates a positive image, and drives economic and social development in the region. Visitors' satisfaction, repeat visits and positive word-of-mouth are encouraged by offering superior products and services compared to similar locations.

6. What is the objective of the draft standard UNE 178510 'Intelligent Tourism Enterprise,' and how does it contribute to the development of smart tourist destinations?

    Answer: UNE 178510 aims to specify requirements and recommendations for Intelligent Tourism Enterprises (ITEs) in terms of management, innovation, technology use, accessibility, and sustainability. It contributes to the development of smart tourist destinations by guiding ITEs towards adopting smart practices and technologies.

References

Abraham R, Schneider J, vom Brocke J (2019) Data governance: a conceptual framework, structured review, and research agenda. Int J Inf Manag 49:424–438. https://doi.org/10.1016/j.ijinfomgt.2019.07.008

Almeida P (2018) Using public Wi-Fi networks to understand tourists' behaviours. Int J Bus Econ Manag Res 4(10):1–7 [PDF]. ijbemr.com

Angele K, Fensel D, Huaman Quispe E, Kärle A, Panasiuk O, Şimşek U, Toma I (2020) Semantic web empowered e-tourism. In: Handbook of e-tourism. Springer. https://doi.org/10.1007/978-3-030-05324-6_66-1

Buhalis D, Leung R (2018) Smart hospitality—interconnectivity and interoperability towards an ecosystem. Int J Hospitality Manag 71:41–50. https://doi.org/10.1016/j.ijhm.2017.11.011

Dogra J, Kale SS (2020) Network analysis of destination management organization smart tourism ecosystem (STE) for E-branding and marketing of tourism destinations. In: Ramos C, Almeida C, Fernandes P (eds) Handbook of research on social media applications for the tourism and hospitality sector. IGI Global, pp 1–16. https://www.igi-global.com/gateway/chapter/246367

Ivars-Baidal JA, Celdrán-Bernabeu MA, Mazón JN, Perles-Ivars ÁF (2021a) Measuring the progress of smart destinations: the use of indicators as a management tool. J Destination Mark Manag 19:100531. https://doi.org/10.1016/j.jdmm.2020.100531

Ivars-Baidal JA, Vera-Rebollo JF, Perles-Ivars ÁF, Femenia-Serra F (2021b) Sustainable tourism indicators: what's new within the smart city/destination approach? J Sustain Tourism 29(2–3):237–255. https://doi.org/10.1080/09669582.2021.1876075

Ji X, Chen J, Zhang H (2024) Smart city construction empowers tourism: mechanism analysis and spatial spillover effects. Humanit Soc Sci Commun 11:1210. https://doi.org/10.1057/s41599-024-03626-w

SEGITTUR (2015) Libro Blanco: Destinos Turísticos Inteligentes. Construyendo el futuro. SEGITTUR. https://www.segittur.es/wp-content/uploads/2019/11/Libro-Blanco-Destinos-Tursticos-Inteligentes.pdf

Shrestha D, Wenan T, Rajkarnikar N, Shrestha D, Jeong S-R (2021) Study and evaluation of tourism websites based on user perspective. J Internet Comput Serv 22(4):65–82. https://doi.org/10.7472/JKSII.2021.22.4.65

UNE (2021) CTN 178 – Ciudades inteligentes (página del comité). Asociación Española de Normalización (UNE). https://www.une.org/encuentra-tu-norma/comites-tecnicos-de-normalizacion/comite?c=CTN+178

Entities

SEGITTUR: https://www.segittur.es/en/
UNE, Spanish Association for Standardization: https://www.en.une.org/
U4SSC, United for Smart Sustainable Cities: https://u4ssc.itu.int/
FEMP, Spanish Federation of Municipalities and Provinces: https://www.femp.es/
RECI, Spanish Network of Smart Cities: https://reddeciudadesinteligentes.es/en/about-us/
OASC, Open & Agile Smart Cities & Communities (OASC): https://oascities.org/
ITU-T SG 20, Internet of things (IoT) and smart cities and communities (SC&C): https://www.itu.int/en/ITU-T/studygroups/2022-2024/20/Pages/default.aspx

Documents and multimedia:

ITU Webinar series on Digital transformation. Episode #31: Digital tourism: bridging the gap between communities and destinations. https://www.itu.int/cities/standards4dt/ep31/
ITU Webinar series on Digital transformation. Episode #19: Tourism in smart cities: Reimagining the road to digital tourism. https://www.itu.int/cities/standards4dt/ep19/
ITU Webinar series on Digital transformation. Episode #15: Smart city platforms for anintegrated management in smart sustainable cities. https://www.itu.int/cities/standards4dt/ep15/
ITU Webinar series on Digital transformation. Episode #6: Smart City Platforms. https://www.itu.int/cities/standards4dt/ep6/United for Smart Sustainable Cities (U4SSC): https://u4ssc.itu.int/
Spanish Smart Cities Model. Ministry of Energy, Tourism and Digital Agenda. April 2017. https://plantl.mineco.gob.es/digital-agenda/Documents/Spanish-Smart-Cities-Model.pdf
Search for Spanish Standards: https://www.en.une.org/encuentra-tu-norma/busca-tu-norma

Smart City Standardization in Japan

David N. Nguyen, Yasuhiro Okuda, and Takeshi Furuno

Abstract

Facing urban agglomeration trends, an ageing population, and disaster risks, Japan has pioneered smart community standards for global use. These standards assist community managers, businesses, and other organizations in strengthening their adaptive capacities. Japan's experience in smart community infrastructure has led to key contributions in ISO, resulting in international standards for sustainable city development. This chapter details the history of ISO standardization and Japan's role in creating deliverables that supporting the UN's Sustainable Development Goals and promotes disaster resilient communities.

Keywords

Smart cities · Japan · Infrastructure · Disaster risk reduction · Sustainability · Disasters

---

D. N. Nguyen (✉)
Tohoku University's International Research Institute for Disaster Science, Sendai, Japan
e-mail: nguyen.david.ngoc.a6@tohoku.ac.jp

Y. Okuda
Disaster Prevention Business Unit at the IMV Corporation, Osaka, Japan
e-mail: y-okuda@imv-corp.com

T. Furuno
Systems and International Standards Development Unit at Japanese Standards Association, Minato, Japan
e-mail: furuno@jsa.or.jp

© The Author(s), under exclusive license to Springer Nature Switzerland AG 2025
L. Anthopoulos (ed.), *Smart City Standardization*, Synthesis Lectures on Computer Science, https://doi.org/10.1007/978-3-031-95959-2_8

8.1 Introduction

Standards play a crucial role for governments, industries, and other organizations. They provide a set of guidelines that businesses can use to build customer confidence, meet regulatory requirements, reduce costs, and gain global market access. For governments and policymakers, standards facilitate global trade, stimulate solutions to national and international issues, reduce costs and expenses, and address policy challenges. Following the devastation of World War II, ISO (the International Organization for Standardization) was founded in 1947 to aid in the reconstruction of affected communities, with its headquarters in Switzerland. Since then, ISO has become the world's leading standardization body, comprising 168 national standardization bodies often represented by national governments, academia, and the private sector.

Globally, growing urbanization has increased pressures on communities to manage their resources effectively to meet local demand. Additionally, climate change and rising disaster risks place further burdens on communities and their managers. To better manage limited financial, human, and material resources, there is a growing global interest in smart city technologies, which offer potential solutions for community managers and other stakeholders.

This paper delves into Japan's development of smart city standards to confront the complex challenges affecting its diverse communities. Specifically, it investigates the historical progression of standardization activities pertaining to smart cities in Japan, with a particular emphasis on the outcomes of ISO Technical Committee (TC) 268, Sub-Committee (SC) 1. Beginning with an examination of ISO's standardization evolution, the chapter subsequently offers a concise overview of publications originating from ISO/TC 268/SC 1, spanning smart community metrics and the formulation of disaster risk reduction strategies. Ultimately, this chapter serves as an introductory exploration of Japan's efforts to disseminate knowledge essential for fostering sustainable smart cities, enhancing community resilience, and advancing the United Nation's Sustainable Development Goals (SDGs) (United Nations 2021).

8.2 Background

Smart city initiatives gained global momentum around 2010, with increased engagement from European countries and Japan. In numerous European nations and the United States, the prevailing smart city model revolved around leveraging information and communication technologies (ICT) to enhance daily life convenience. In contrast, Japan and other Asian countries prioritized an environment-centric approach, emphasizing initiatives like improving energy efficiency within urban areas. This shift towards eco-friendly urban development spurred an international push for standardizing smart city indicators. In October 2011, separate proposals from Japan, France, and Canada were submitted to

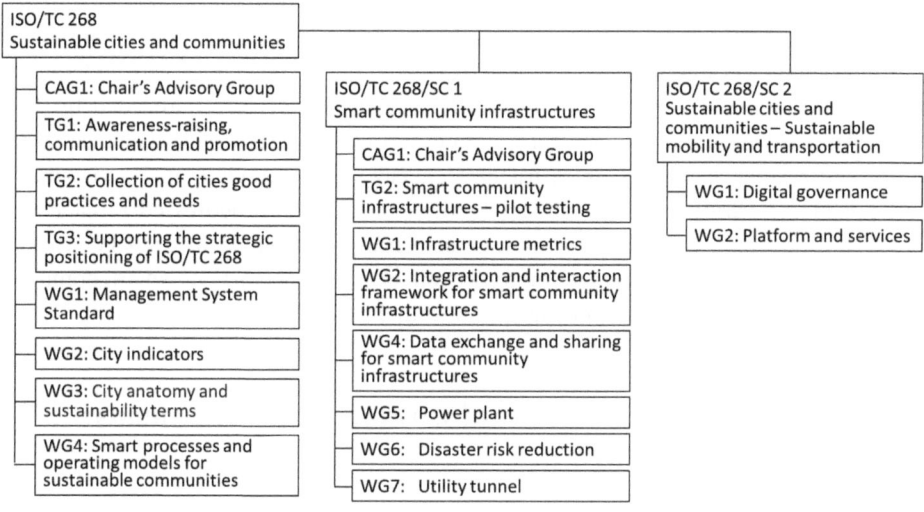

Fig. 8.1 The structure of ISO/TC 268 as of June 2024

ISO, each focusing on developing deliverables for city and community indicators. ISO's Technical Management Board responded by establishing Technical Committee 268 (TC 268) in February 2012, tasked with overseeing these standardization efforts. Concurrently, Working Group 1 (WG 1) under ISO/TC 268 was designated to develop a management system standard for sustainable development, while Subcommittee 1 (SC 1) was formed to address smart community infrastructure standardization, following Japan's proposal. Furthermore, the International Electrotechnical Commission (IEC) and the International Telecommunication Union (ITU) also embarked on standardization endeavors related to cities, particularly smart cities (International Electrotechnical Commission 2020). Subsequent to the establishment of ISO/TC 268, additional Working Groups were formed under its umbrella, including under ISO/TC 268/SC 1. Notably, in 2022, the scope of ISO/TC 268/SC 1/WG 3 expanded as it transitioned to Subcommittee 2 (SC 2) under ISO/TC 268, focusing on sustainable mobility and transportation (Fig. 8.1).

8.3 Research Methodology: A Case Study of Japan

Communities in Japan face multiple demographics challenges. For example, the continuing agglomeration in major cities such as the Greater Tokyo area which presents numerous challenges due to overcrowding. In contrast, Japan also has many communities that are facing issues of depopulation and ageing. Finally, due to the country's geographic proximity to various natural hazards, many communities are exposed to potential disaster risk. The challenges facing Japanese communities has pushed the country to utilize its experiences to develop smart city standards that could be adopted by global communities.

In this study, a comprehensive literature review of ISO deliverables was undertaken to chart the evolution of smart community infrastructure standards spearheaded by Japan. These standards provide guidance for community managers striving to enhance sustainability and resilience through infrastructure endeavors. The review analyzed published works spanning various periods, dissecting the central themes addressed in each deliverable. By exploring the evolution of ISO deliverables produced by ISO/TC 268/SC1, readers can develop a nuanced comprehension of the trajectory of smart community infrastructure standards, from their inception to their specialized applications.

8.3.1 Strategic Level: A Review of Existing Smart Community Metrics and Priorities

ISO/TC 268/SC 1/WG1 first conducted a survey of examples of existing relevant activities and reviewed literature to identify the directions for metrics of smart community infrastructures that can help contribute to improving the sustainability of communities. This process considered the different needs of various communities and the geographic and economic diversity in respect to global diversity. The results of the survey and analysis were summarized in the technical report ISO/TR 37150, Smart community infrastructures—Review of existing activities relevant to metrics, which was published in 2014 (International Organization for Standardization 2014).

In evaluating sustainable development, the degree of contribution to sustainability in the economic, environmental, and social aspects is generally used as the axis of evaluation (International Organization for Standardization 2019). In the review and discussion of metrics for smart community infrastructure in ISO/TC 268/SC 1/WG 1, "social aspect" was replaced by "residents' perspective" and "economic aspect" by "community managers' perspective," including government officials and infrastructure operators, to better clarify the perspective on infrastructure. In reviewing examples of activities related to metrics for community infrastructure, a functional model of a city that consists of three layers was used, with a service layer at the top, a facility layer where services are implemented second, and third, an infrastructure layer that supports these two layers.

To contribute to sustainable development, ISO/TR 37150 identified the following desirable features for smart community infrastructure metrics:

- be harmonized for every stakeholder, be useful for infrastructure transactions, facilitate evaluation of technical performance, be applicable to various stages of development of infrastructure, and reflect dynamic characteristics.
- in aggregate, address multiple issues/tradeoffs and consider advanced features such as interoperability and efficiency.

- be applicable to diverse communities (e.g., geographical location, sizes, level of economic development).
- consider multiple community infrastructures (e.g., energy, ICT), provide technologically implementable solutions, and allow a holistic perspective of multiple community infrastructures.
- allow evaluation of the technical performance of community infrastructures and be based on transparent and scientific logic.

In summary, smart community infrastructure metrics are defined as a measurement or quantification method and scale of the technical performance of community infrastructures which allow a holistic perspective of multiple infrastructures in communities, have dynamic properties, takes into account the long-term aspects of communities, and enable understanding of the diversity of communities.

The gap analysis in ISO/TR 37150 revealed that none of the existing activities had all the desirable features of smart community infrastructure metrics, and it was recommended to develop a set of new general principles and requirements for smart community infrastructure metrics, taking into consideration the lessons learned from relevant activities.

8.3.2 Process Level: Identifying Principles and Requirements for Smart Community Infrastructure and the Development of Relevant Frameworks

8.3.2.1 The Development of Principles and Requirements for Smart Community Metrics

To evaluate community infrastructure from a cross-sectional perspective, an approach was considered that first defines the smartness of community infrastructure, then establishes basic rules, and finally to develop smart community infrastructure metrics based on these rules.

ISO/TS 37151 established principles for defining, identifying, optimizing and harmonizing smart community infrastructure performance metrics and specified requirements for procedures for identifying such performance metrics (International Organization for Standardization 2015). Following ISO/TS 37151, the expansion of themes for standardization in three directions has been underway in ISO/TC 268/SC 1.

The first direction is standardization to support continuous improvement of community infrastructures. To promote continuous improvement, ISO/TC 268/SC 1/WG 1 developed a standard to provide a systematic framework for assessment, the CIMM (community infrastructure maturity model), which includes the five reference levels of maturity in each of the characteristics of the community infrastructure. The model is intended to serve as a tool to measure the current maturity of community infrastructure relative to desired future improvements. In addition, a set of standards has been developed in ISO/TC 268/SC 1/WG 2 to provide a framework for evaluation of infrastructure integration and evaluation of interactions to encourage each infrastructure to be organized as part of a smart city and to coordinate interactions among multiple smart community infrastructures in the development, operation, and maintenance of a smart city.

The second direction is standardization of smartness specific to individual infrastructures. WG 3 (Smart transportation), WG 4 (Data exchange and sharing for smart community infrastructures), WG 5 (Power plant), and WG 7 (Utility tunnel) have been established under ISO/TC 268/SC 1 and multiple standards for individual infrastructures have been developed, including a governance framework, guidelines for smart operation and maintenance, and methods for evaluating the performance. As a result of the many standard development efforts in ISO/TC 268/SC 1/WG 3 on a wide range of transportation infrastructure-related topics, SC 2 was established under ISO/TC 268 in 2021 to carry on and develop the themes of ISO/TC 268/SC 1/WG 3.

The third direction is standardization of the use of smart community infrastructures for disaster risk reduction. To this end, ISO/TC 268/SC 1/WG 6 (Disaster risk reduction) was established in 2020 to develop principles, guidelines, and requirements for the implementation of community infrastructures that are useful for addressing critical needs for disaster risk reduction, with one Technical Report and one International Standard published and two International Standards under development as of June 2024.

ISO/TS 37151 Smart community infrastructures—Principles and requirements for performance metrics, developed by ISO/TC 268/SC 1/WG 1 and published in May 2015, specifies the principles for defining, identifying, optimizing, and harmonizing performance metrics for smart community infrastructure in Clause 5, and requirements for a common approach to identifying performance metrics in Clause 6. It has been decided to convert this Technical Specification to an International Standard, and its publication is expected in 2024.

In Clause 5, the ideal properties are introduced, which should be considered in defining or identifying a set of community infrastructure performance metrics. It is recommended that community infrastructure performance characteristics should be related to community issues or priorities. It is required that the interests of stakeholders should be considered in a balanced manner that encompasses the multiple perspectives of different stakeholders in the community when identifying smart community infrastructure performance metrics.

In Clause 6, a step-wise approach is specified as requirements to identify smart community infrastructure performance metrics. The 1st step is understanding the perspectives of key stakeholders for community infrastructures, the 2nd step is identifying the needs from the perspective of the key stakeholders, the 3rd step is translating the needs into performance characteristics, finally the 4th step is identifying metrics appropriate for measuring those performance characteristics.

The list of ideal properties of smart community infrastructure performance metrics, introduced in Clause 5, identifies basic properties that should be considered in identifying smart community infrastructure performance metrics. For example, they should:

- include items that help trades of community infrastructure products and services;
- facilitate evaluation of the technical performance of community infrastructures that contributes to enhance sustainability and resilience of communities;
- consider synergies and trade-offs arising from multiple issues or aspects a community faces, such as environmental impacts and quality of community services;
- consider advanced features of community infrastructures such as interoperability, expandability, and efficiency;
- be applicable to a diverse range of communities with different sizes and levels of economic development, as well as to individuals within communities with varying age, gender, income, disability, ethnicity;
- consider multiple community infrastructures that support operations and activities of communities, including energy, water, sewer, transportation, waste, and ICT;
- consider an integrated system which includes the interaction and coordination of multiple community infrastructures.

Fourteen needs are identified in Clause 6, which are considered the minimum requirements. They are five needs from the residents' perspective (availability, accessibility, affordability, safety and security, and quality of service), five needs from the community managers' perspective (operational efficiency, economic efficiency, availability of performance information, maintainability, and resilience), and four needs from the environmental perspective (effective use of resources, mitigation of climate change, prevention of pollution, and conservation of ecosystem).

To identify smart community infrastructure performance metrics with an approach that follows the provisions of Clause 5 and Clause 6, it is necessary to consider critical needs, including those minimal set of needs, from the perspective of three stakeholders: residents, community managers, and the environment.

Examples are provided in Annex A of ISO/TS 37151, which demonstrate the validity of the stepwise approach specified in Clause 6 for existing metrics or key performance indicators for various community infrastructures in different countries, and they can be helpful to users of the standard.

8.3.2.2 A Common Framework for Development and Operation of Smart Community Infrastructure

ISO/TR 37152 Smart community infrastructure—Common framework for development and operation, developed by ISO/TC 268/SC 1/WG 2 and published in 2016, presents findings on methodologies for planning, developing, operating, and maintaining community infrastructure to promote harmonization and ensure that interactions between multiple infrastructures are properly coordinated (International Organization for Standardization 2016). In today's smart cities, where multiple infrastructures are becoming networked, various interactions between infrastructures may occur, resulting in an increase in previously unforeseen risks. These risks are due to factors such as, for example, lack of consistency among infrastructures, increased difficulty in predicting and controlling the impact of one infrastructure on another, and increased difficulty in integrated management in the development and operation of multiple infrastructures.

To reduce these risks, this standard provides a framework, or process and methodology, for managing smart community infrastructure interactions and incorporating appropriate measures into planning and operation. Adopting this framework (processes and methodologies) will help each stakeholder to ensure efficient and effective development and operation when participating in the development and operation of community infrastructure.

ISO/TR 37152 summarizes the findings of a study of a framework for addressing risks in infrastructure development and operation in smart cities, where multiple infrastructures are networked together, as consisting of the following three elements:

Element A is the allocation of specifications to each component and validation of the allocating procedures to ensure consistency among infrastructures.

Element B is the specification requirements related to interaction between outside/inside smart community infrastructures and adoption of appropriate measures into planning and operation.

Element C is the adoption of processes to facilitate information sharing and communication among stakeholders.

Based on the findings of a study described in ISO/TR 37152, ISO 37155-1 Framework for integration and operation of smart community infrastructures—Part 1: Recommendations for considering opportunities and challenges from interactions in smart community infrastructures from relevant aspects through the life cycle, developed in 2016 by ISO/TC 268/SC 1/WG 2, provides guidelines for specifications to manage smart community infrastructure interactions and incorporate appropriate measures in planning and operation during the planning, development, operation, and maintenance stages to realize solutions that are corresponding to element B (International Organization for Standardization 2020a).

ISO 37155-2 Framework for integration and operation of smart community infrastructures—Part 2: Holistic approach and the strategy for development, operation and maintenance of smart community infrastructures, developed in 2021 by ISO/TC 268/SC 1/WG 2, provides guidelines for assigning consistent specification requirements to each component of a smart community infrastructure system at the planning and design phases and for validating the assignment procedures in order to realize solutions that are corresponding to element A (International Organization for Standardization 2021). It also provides guidelines for specifications to adopt appropriate measures for planning and operation in order to properly implement verification and validation during the operation and maintenance phases.

8.3.2.3 Smart Community Infrastructure Maturity Model for Assessment and Improvement

ISO 37153 Smart community infrastructures—Maturity model for assessment and improvement, developed by ISO/TC 268/SC 1/WG 1 and published in 2017, describes a community infrastructure maturity model (CIMM) and a standardized approach for the assessment and improvement utilizing CIMM (International Organization for Standardization 2017).

ISO 37153 combines the five-level rating in the CIMM, the commonly used capability maturity model, with the critical needs from the three stakeholder perspectives (residents, community managers, and the environment) that should be considered for community infrastructures. It provides tools to identify gaps in the current level of community infrastructure performance, assists in understanding performance levels, processes, interoperability of community infrastructures, their contribution to the city, and assists in setting improvement targets to guide investments. An achievement criteria table (ACT) was developed and used to assess technical aspects of community infrastructures such as performance, processes, and interoperability, as well as to assess contribution of community infrastructures to community-wide priorities. The achievement criteria table consists of the following components: characteristics, objectives of each characteristic, five levels of characteristics, and descriptions or definitions of criteria for those characteristics which define each level.

The assessment of community infrastructures uses two aspects: the technical assessment is useful for operators, supervisory authorities and community infrastructure vendors, while the contribution assessment is useful for government decision makers and development agencies.

ISO 37153 provides the following procedure for the preparation of the ACT:

- determine the purpose, assessment aspect and target community infrastructure
- select characteristics that fit the purpose
- define the characteristics for each of the five maturity levels in accordance with the CIMMs
- organize the characteristics and the maturity levels into criteria in a table.

To determine the purpose of the assessment, users should determine the target community, identify and prioritize community issues, and identify and analyze relevant community infrastructure and its impacts on the prioritized community issues. For a contribution assessment, users should refer to the UN SDGs or national development plans to prioritize community issues and to select community-wide characteristics.

The selection of characteristics for the technical assessment of community infrastructures, as shown in ISO/TS 37151, calls for understanding the perspectives of key stakeholders of community infrastructure, identifying the needs that are important to their perspectives including consideration for 14 critical needs, and then translating these needs into performance characteristics. Table 8.1 shows a partial example of the ACT for electric power supply infrastructure.

In Table 8.1, the left-most column consists of characteristics derived from the 14 needs identified in ISO/TS 37151. Based on the needs associated with the characteristics, a description of the criteria for the five maturity levels of infrastructure under consideration, in terms of performance, operation, and relationship to other infrastructure, is provided under the headings "Level 1" through "Level 5" in the table. The ACT can be used to identify gaps between the current status of the infrastructure and its goals, and to help determine measures for improvement. It is also possible to combine multiple solutions found in the ACT to identify measures to improve maturity through coordination among infrastructures.

ISO 37153 aims to facilitate cities in assessing the level of their infrastructure, set targets and areas for improvement, and set policies for progressively more mature levels of community infrastructure. This international standardization of a framework to support the gradual improvement of smart community infrastructure is expected to have the effect of supporting the continual improvement of smart community infrastructure promoted by various levels of government in different countries toward the achievement of SDG 11 (Make cities and human settlements inclusive, safe, resilient, and sustainable).

8.3.3 Technical: The Development of Deliverables for Disaster Risk Reduction

8.3.3.1 The Emergence of Smart Community Infrastructure for Disaster Risk Reduction

In order to protect communities against natural hazard risks, infrastructures can play a key role in strengthening resilience. The approach of using smart community infrastructures for disaster risk reduction is essential for smart cities to adapt to climate change and realize a sustainable and resilient society. In addition, the Sendai Framework for Disaster Risk Reduction 2015–2030 (United Nations 2015), which was adopted as the outcome document of the Third United Nations World Conference on Disaster Risk Reduction in 2015, identifies four priority areas to guide the focused actions by states, regional and international organizations and other relevant stakeholders: Understanding disaster risk; Strengthening disaster risk governance to manage disaster risk; Investing in disaster risk reduction for resilience; and Enhancing disaster preparedness for effective response and to "Build Back Better" in recovery, rehabilitation and reconstruction.

8 Smart City Standardization in Japan

Table 8.1 Example of achievement criteria table for electric power supply infrastructure

Characteristics	Needs	Objectives	Metrics	Level1	Level2	Level3	Level4	Level5
Power system quality level	Availability	To identify quality levels of the power system, including its management and operation	Average interruption duration per year per customer	***	***	***	***	***
	Stability							
	Redundancy							
Efficiency of management and operations for keeping the balance	Operational efficiency	To identify the level of efficiency of management and operations for keeping a balance between supply and demand through tie-lines among different regions	Amounts of interconnected power exchange by the tie-lines	***	***	***	***	***
	Interoperability							

In light of this situation, there is a move to develop an international standard for principles and basic requirements for disaster-resistant infrastructure that provides a basic framework for disaster prevention using community infrastructure to support local governments and infrastructure service providers seeking to strengthen community resilience through infrastructure investment, stockpiling, and disaster prevention training. Based on Japan's experience in disaster response, efforts are underway to develop international standards for various technologies for disaster risk reduction and services that support the framework for disaster risk reduction.

8.3.3.2 Survey Results and Analysis of the Current State of Infrastructure for Disaster Risk Reduction

WG 6 (Disaster Risk Reduction) was established under ISO/TC 268/SC 1 in October 2020, based on a proposal from Japan, to develop deliverables regarding the use of smart community infrastructures that can contribute towards disaster risk reduction. The deliverables produced in WG6, through disaster risk reduction, can contribute to achieving targets related to making cities resilient under SDG 11, such as climate change adaptation and planning and implementing disaster preparedness.

ISO/TR 6030 Smart community infrastructures—Disaster risk reduction—Survey results and gap analysis, was developed by WG6 and published in 2022. This technical report provides a summary of infrastructure utilized for disaster risk reduction throughout the world, and categorizes them based on type of disaster focus, infrastructure type, infrastructure functions, and target areas for utilization (International Organization for Standardization 2022).

The technical report identifies that ICT infrastructure accounts for nearly half of the results based on a literature review, and that these technologies play a key role in various phases of a disaster, in order to provide disaster management related information. This includes the use of sensor technologies to collect hazard information, mobile communication systems that can transmit data and information, and the use of satellite systems.

For the built environment, the report provides examples of hazard resistant infrastructure, as well as examples on improvements to existing designs that could reduce hazard risks. Other examples include technologies that can monitor building conditions to inform stakeholders about the health of the structure and when appropriate action should be taken.

For transportation infrastructure, examples of disaster risk reduction technologies include the use of seismometer systems for high-speed railway networks, flood warning systems for vehicles, and early warning systems for trains.

For energy infrastructure, the importance of continuous operation of various energy production and management facilities was reported. Also addressed was the importance in establishing food security which plays a key role in the recovery and Build Back Better process.

Based on the results of the surveys and gap analyses, the following areas are identified as possible topics for standardization:

1. A common framework for the implementation of disaster risk reduction in smart community infrastructure
2. Governance systems (e.g., Local multi-stakeholder dialogue framework, Community engagement and participation framework, Risk financing systems, Disaster information sharing systems, Intergovernmental data sharing framework, Risk assessment systems)
3. Prevention systems (e.g., Hazard identification systems, Hazard monitoring, detect, and prediction systems, Disaster mitigation systems)
4. Preparedness systems (e.g., Food security systems, Energy security systems)
5. Response systems (e.g., Evacuation systems, Drones for rescue, emergency, and firefighting use, Emergency communication systems)
6. Build Back Better systems (e.g., Food distribution systems, Smart and resilient buildings, Public spaces, Evaluation and auditing systems)

ISO/TR 6030 has summarized examples of community infrastructures for disaster risk reduction and it is expected that the possible areas for standardization, identified from the gap analysis with needs and expectations, can become a basis for future consideration of standardization themes for community infrastructure that can contribute to disaster risk reduction.

8.3.3.3 Guidance for Implementation of Seismometer Systems

The second project in ISO/TC 268/SC 1/WG 6 is the development of ISO 37174 Smart community infrastructures—Disaster risk reduction—Guidance for implementing seismometer systems, which specifies an application-based classification of seismometer systems and was published in February 2024 (International Organization for Standardization 2024c).

Earthquakes are one of the most devastating of all the natural hazards because of their impact. Effective use of seismometer systems will contribute to preventing and reducing damage caused by earthquakes, and to maintain the community's level of services and quality of life after the earthquakes by enabling more informed emergency responses. Data from seismometer systems also improves the understanding and modelling of ground motion and structural behaviour, leading to improved seismic design regulations and improved seismic risk modelling.

In seismically active countries, by installing appropriate seismometers, the data can be used not only for land use control, structural design of buildings and other facilities, but also for emergency responses, evacuation guidance, and the development of business continuity plans.

ISO 37174 aims to assist stakeholders at various levels of governments (including developed and emerging and developing countries), planners, developers, and operators of the communities, to optimize their investments in community development by deploying and utilizing seismometer systems as a tool for disaster risk reduction from earthquakes. This document also describes a categorization of the purposes of seismometer systems for achieving disaster risk reduction. Analysis of the data obtained from seismometer systems provides information for managing risk and reducing the impact on people, organizations, infrastructures, and livelihoods as well as for planning preventive measures and emergency responses after an earthquake. For these reasons, effective use of this data will enable smart communities to enhance their resilience to earthquakes.

ISO 37174 classifies the purposes for which seismometer systems are used into categories A through J (e.g., A for hazardous area survey, H for emergency stoppage in the event of an earthquake, I for structural damage survey) and subcategory +L for long-period seismic motion measurement which can damage structures such as high-rise buildings and oil tanks. The guidance allows the user to select seismometer systems according to the use cases. After selecting the purpose for which seismometer systems are to be used, the user can select the appropriate seismometer by checking the necessary data specifications from the selected purpose. By selecting and using the appropriate seismometer systems, damage caused by earthquakes can be reduced.

The development of guidance on how to select seismometer systems was proposed to ISO by Japan in 2023 and the proposal was approved in February 2024. The development of ISO 37194 Smart community infrastructures—Disaster risk reduction—Guidance for the process of selecting the seismometer systems for specific purposes is the fourth project in ISO/TC 268/SC 1/WG 6. The document specifies how to select seismometer systems through examples of the selection and use of seismometer systems based on the categories defined in ISO 37174 with the intention to help implementers and suppliers of seismometer systems. ISO 37194 is in Working Draft (WD) stage as of June 2024 (International Organization for Standardization 2024a).

8.3.4 Cross Organizational: A Basic Framework for Implementation of Smart Community Infrastructure for Disaster Risk Reduction

The third project in ISO/TC 268/SC 1/WG 6 is the development of ISO 37179 Smart Community Infrastructures—Disaster risk reduction—Basic framework for implementation, which specifies principles and general requirements for the implementation of smart community infrastructures to realize the concept of disaster risk reduction, and is in DIS stage as of June 2024 (International Organization for Standardization 2024b).

About hydrometeorological and environmental hazards, climate change is often a multiplier of disaster risk, as it is projected to exacerbate existing hazard risks through increased frequency or intensity. For this reason, it is essential for smart communities to design DRR measures that allow communities to adapt to climate change and become sustainable and resilient.

It is also essential to plan, build, utilize, maintain and improve community infrastructures considering disaster risk. Such infrastructure can also be utilized alongside existing community infrastructure and supplemented by nature-based solutions.

The Sendai Framework for Disaster Risk Reduction identifies several key areas on how infrastructure could be used to reduce disaster risk and strengthen community resilience to shocks caused by natural hazard that may lead to infrastructure service disruptions. This includes investing in disaster risk reduction, the adoption of infrastructure and technologies that can identify hazard risk, and technologies that facilitate the sharing of information which could be used for disaster risk reduction services for community managers.

ISO 37179 establishes ten principles, with general requirements for each principle, which contributes to the achievement of the four priorities for actions of the Sendai Framework for Disaster Risk Reduction, with regards to community infrastructure. These principles include four overarching principles and six principles for focus areas for the continuous improvement of disaster risk reduction. Together, these ten principles provide community stakeholders a framework to leverage smart community infrastructure to increase community resilience, both in the development of new community infrastructure and in the retrofitting of existing infrastructure.

8.3.5 Discussion

A comprehensive review of the development of ISO deliverables in ISO TC 268 SC1 highlights the significant evolution of standards for smart community infrastructure. Initially, these standards were broad and conceptual, incorporating holistic frameworks and general requirements. Over time, the focus narrowed to more specific areas such as disaster risk reduction and the use of technologies like seismometers. ISO TR 6030, which surveyed and analyzed real-world examples of DRR infrastructure, identified how standards could draw on global examples of smart community infrastructure to develop new standards that promote organizational collaboration in planning. This culminated in the development of ISO 37179, which emphasizes cross-organizational aspects by identifying key stakeholders in smart community infrastructure planning for disaster risk reduction and defining their roles.

8.4 Conclusion

As members of the global community, we live in a world where change is constant and disruptions can be wide-reaching. In order to anticipate and adapt to these changes, standards can provide guidelines that can be used by policy makers and enterprises, for the benefit of multiple community stakeholders. Facing numerous demographic pressures and constant exposure to disaster risk, Japan has accumulated significant experience in the development of smart city infrastructure which could be used by community managers to adapt to these changes. These experiences have led to Japan's participation in ISO to develop various deliverables that could be used by community managers as a guideline in the development, implementation, and monitoring of these infrastructures. Due to Japan's extensive experience with earthquakes, typhoons and other disasters, there has been active efforts to share them to the world in order to contribute to the enhancement of disaster prevention and mitigation capabilities in various communities.

Under ISO/TC 268/SC 1, international standards on principles and requirements for smart community infrastructure performance metrics have been developed under the leadership of Japan, and a series of standards have been developed for various examples of best practices in the field of smart transportation, which help increase the contribution of smart community infrastructures to sustainable development.

With regard to developments in the energy infrastructure field, WG 5 (Power plant) was established under ISO/TC 268/SC 1 in October 2017, based on a proposal from Japan, to develop international standards for smart power infrastructure. ISO 37160:2020 Smart community infrastructure—Electric power infrastructure—Measurement methods for the quality of thermal power infrastructure and requirements for plant operations and management, developed by WG6 and published in 2022. ISO 37160 aims to realize a sufficient and stable power supply while taking into consideration the reduction of environment impact and improving sustainability of communities through the stable and efficient operation of thermal power infrastructure over its entire life cycle (International Organization for Standardization 2020b). In particular, it helps developing countries to select, build, and operate smart power infrastructure when they seek to introduce cost-effective thermal power infrastructure for industrial development.

As of June 2024, demonstrations of smart city development aiming at realizing energy-saving, CO_2-saving, safe, secure, and sustainable smart cities are actively ongoing in Japan. Through the cooperation between companies and local governments, the experience obtained from these demonstrations is being fed back into the development of smart city-related standards. As the ageing population continues to grow, the importance of community infrastructure to support social systems in times of disaster and in daily life increases. The challenge will be to develop international standards that contribute to the realization of smarter community infrastructure that takes into account needs for the safe and healthy lives of a wide range of physically, mentally, and cognitively diverse residents.

8.5 Revision Questions with Answers

What are International Standards?
International Standards are guidelines, specifications, criteria, or best practices developed and published by International Organization for Standardization (ISO). These standards are designed to ensure the quality, safety, efficiency, and interoperability of products, services, and systems across different countries.

Why was ISO established?
ISO was established in 1947 to create and publish international standards across a wide range of industries. Much of its impetus stemmed from the reconstruction needs following World War II, as many communities were in dire need of rebuilding and reformation guidelines. International standards produced by ISO can facilitate international trade, improve industry practices, and protect consumers and the environment by providing consistent and universally accepted criteria.

What are the problems facing Japanese communities?
Japan faces several major challenges affecting its communities. Urban centers such as the greater Tokyo and Osaka areas continue to experience agglomeration which leads to pressure on infrastructure, housing and services. At the same time, many rural regions in the country are experiencing continued depopulation and ageing, which creates challenges in maintaining public services and infrastructure. Both communities however, are also at risk of disasters due to Japan's exposure to many natural hazards due its geological and hydrometeorological setting.

What are smart community infrastructures?
Smart city infrastructures are referred to as smart community infrastructures in ISO, in order to better reflect the broader categories of human settlements in the world. Smart community infrastructures are community infrastructure with enhanced technological performance that is designed, operated and maintained to contribute to sustainable development and resilience.

How can international standards for smart community infrastructures help communities?
International standards for smart community infrastructure can provide numerous benefits for communities such as consistency, interoperability, improved safety and resilience, economic benefits, improved quality of life, technological advancement, knowledge sharing, and environmental sustainability.

Acknowledgements We would like to express our sincere gratitude to the dozens of international experts that have worked with us in the development of various ISO deliverables over these past several years. Through their invaluable guidance and support, the standards that are published by the various committees, can contribute to the sustainability and resiliency of global communities. In addition, we would like to thank our colleagues at the Japanese Standard's Association and Tohoku University for their continued support.

References

International Electrotechnical Commission (2020) IEC 63152:2020: Smart cities—city service continuity against disasters—the role of the electrical supply. https://webstore.iec.ch/publication/60486

International Organization for Standardization (2014) ISO/TR 37150:2014: Smart community infrastructures—review of existing activities relevant to metrics

International Organization for Standardization (2015) ISO/TS 37151:2015: Smart community infrastructures—principles and requirements for performance metrics. https://www.iso.org/standard/61057.html

International Organization for Standardization (2016) ISO/TR 37152:2016: Smart community infrastructures—common framework for development and operation. https://www.iso.org/standard/66898.html

International Organization for Standardization (2017) ISO 37153:2017: Smart community infrastructures—maturity model for assessment and improvement

International Organization for Standardization (2019) ISO guide 82:2019: guidelines for addressing sustainability in standards. https://www.iso.org/standard/76561.html

International Organization for Standardization (2020a) ISO 37155-1:2020: Framework for integration and operation of smart community infrastructures. Part 1: Recommendations for considering opportunities and challenges from interactions in smart community infrastructures from relevant aspects through the life cy. https://www.iso.org/standard/69241.html

International Organization for Standardization (2020b) ISO 37160:2020: Smart community infrastructure—electric power infrastructure—measurement methods for the quality of thermal power infrastructure and requirements for plant operations and management

International Organization for Standardization (2021) ISO 37155-2:2021: Framework for integration and operation of smart community infrastructures. Part 2: Holistic approach and the strategy for development, operation and maintenance of smart community infrastructures

International Organization for Standardization (2022) ISO/TR 6030:2022: Smart community infrastructures—disaster risk reduction—survey results and gap analysis

International Organization for Standardization (2024a) ISO/AWI 37194: Smart community infrastructures—disaster risk reduction—guidance for the process of selecting seismometer systems suitable for specific purposes (In Progress)

International Organization for Standardization (2024b) ISO/DIS 37179: Smart community infrastructures—disaster risk reduction—basic framework for the implementation of disaster risk reduction. https://www.iso.org/standard/69265.html

International Organization for Standardization (2024c) ISO 37174:2024: Smart community infrastructures—disaster risk reduction—guidance for implementing seismometer systems

United Nations (2015) Sendai Framework for disaster risk reduction 2015–2030, vol. A/CONF.224. United Nations. http://www.preventionweb.net/files/resolutions/N1509743.pdf

United Nations (2021) The 17 goals. United Nations Department of Economic and Social Affairs. https://sdgs.un.org/goals

Dr. David N. Nguyen graduated from Tohoku University's Department of Civil Engineering. In addition, Dr. Nguyen is also an alumnus of the University of Hawaii and the University of Tokyo. Currently Dr. Nguyen is a Research Fellow at Ritsumeikan Asia Pacific University while holding a concurrent position at Japan's National Research Institute for Earth Science and Disaster Resilience. Dr. Nguyen has experience in authoring several deliverables in ISO.

Mr. Yasuhiro Okuda is an expert employed at the IMV Corporation based in Osaka, Japan. The IMV Corporation is renowned for its environmental simulation and vibration test systems. Due to Mr. Okuda's expertise, he has been involved in a number of seismology related projects in hazard prone regions, such as infrastructure resiliency in China, Indonesia, Japan, among others.

Mr. Takeshi Furuno is a senior expert at the Japanese Standards Association. Mr. Furuno has guided the development of dozens of ISO deliverables across multiple Technical Committees, from Smart community infrastructures, Occupational health and safety management systems, Security and resilience, among many others.

Smart City Standardization in Greece

Panagiotis Karadimos and Leonidas Anthopoulos

Abstract

Modern cities face many problems that affect their residents' quality of life. Key issues include overpopulation and congestion, environmental pollution, lack of green spaces, inadequate waste management, climate change, inadequate infrastructure, social inequality, and public health issues. These problems are expected to intensify in the coming years. Today, cities account for almost 50% of the world's population and generate 80% of the global GDP. These percentages are only expected to rise as more people move into the cities. Organizations such as the International Organization for Standardization (ISO) and the International Telecommunication Union (ITU) develop standards that help cities address the issues they face, promoting sustainable practices and improving the quality of life for their residents. These standards include the ISO 37100 series on sustainable cities and communities (International Organization for Standardization. ISO 37100 series: Sustainable cities and communities. ISO/TC 268). Although sustainable development is a global challenge, strategies to achieve urban sustainability are largely local. Therefore, strategies vary in scope and content from country to country and region to region. To support national strategies and local initiatives in Greece, the Hellenic Organization for Standardization (ELOT) provides Greek versions of the international standards: ELOT ISO 37101, "Sustainable cities—Management systems for sustainable cities– Requirements with guidance for use" (Hellenic Organization for Standardization, ELOT ISO 37101: Sustainable cities—Management

---

P. Karadimos (✉) · L. Anthopoulos
Department of Business Administration, School of Economics and Business, University of Thessaly, Geopolis Campus, Larissa, Greece
e-mail: pkaradimos@uth.gr

L. Anthopoulos
e-mail: lanthopo@uth.gr

systems for sustainable cities—Requirements with guidance, 2018a) and ELOT ISO 37100, "Sustainable cities—Terms and definitions of concepts" (Hellenic Organization for Standardization, ELOT ISO 37100: Sustainable cities—Terms and definitions of concepts, 2018b). These standards help Greek cities meet international obligations and achieve Sustainable Development Goals. Furthermore, to support the sustainability strategies of municipalities, local communities, and stakeholders, and to transform Greek cities into sustainable, smart, resilient, and climate-adaptable ones, the Hellenic Organization for Standardization (ELOT) developed the standard ELOT 1457, "Sustainable development in cities—Indicators for sustainability" (Hellenic Organization for Standardization, ELOT 1457: Sustainable development in cities—Indicators for sustainability, 2018c). This standard through the development of performance indicators covers the essential functions for cities to consistently and inclusively provide quality services, ensuring a high quality of life and an attractive environment for citizens, visitors, and investors. The present study presents a summary of the ELOT 1457 standard as well as ELOT's participation in the development of new standards.

Keywords

Urban sustainability • Sustainable development goals • International standards • ELOT 1457 • Sustainable cities • Smart cities

9.1 Introduction

Urban sustainability is a critical issue that addresses cities' ability to meet the needs of their current residents without jeopardizing the ability of future generations to meet their own needs. Sustainable urban development involves the balanced management of resources and environmental challenges, aiming to create a healthy, enjoyable, and secure environment for cities' inhabitants. The UN Sustainable Development Goal 11 (United Nations 2015) defines sustainable cities as those that are dedicated to achieving green sustainability, social sustainability, and economic sustainability. Green sustainability encompasses the protection of natural resources and ecosystems, the reduction of pollution, and the promotion of biodiversity. Economic sustainability involves efficient resource use, promoting economic growth, creating jobs, and maintaining environmental balance. Social sustainability focuses on ensuring social cohesion, access to essential services (such as health, education, and social welfare), and promoting equality.

In recent years, the concept of the "smart city" has gained significant traction, with many cities around the world actively pursuing this path. But what exactly defines a smart city? If someone looks for a precise definition of a smart city, they will not find a single one but instead encounter various alternatives, leading to ambiguity in its meaning (Anthopoulos 2017). The ISO, "Smart cities Preliminary Report" (International Organization for Standardization 2014) recognizes smart city as a new concept and a new model, that applies the new generation of information technologies, such as the Internet of Things, cloud computing, big data, and space/geographical information integration, to facilitate the planning, construction, management and smart services of cities. Moreover, it defines smart city objectives to pursue: convenience of public services; delicacy of city management; livability of living environment; smartness of infrastructures; and long-term effectiveness of network security. The ITU, "Setting the framework for an Information and Communication Technologies (ICT) architecture of a smart sustainable city" (International Telecommunications Union 2014b) emphasizes on ICTs and considers a smart sustainable city as an innovative city that uses ICTs and other means to improve quality of life, efficiency of urban operation and services, and competitiveness, while ensuring that it meets the needs of present and future generations with respect to economic, social and environmental aspects.

Sustainable cities and smart cities are interconnected through their shared goals of improving urban quality of life while minimizing environmental impact. Sustainable cities focus on resource efficiency and social equity through strategies like renewable energy and green infrastructure. Smart cities use technology and data to optimize urban operations, enhance service delivery, and promote sustainability. Both aim to create resilient, livable cities that meet the needs of current and future generations effectively.

The ISO and the ITU have developed several standards focusing on sustainable cities and communities. These standards provide frameworks and guidelines to help cities and communities to become more sustainable and resilient. More specifically, the International Organization for Standardization (ISO) has a dedicated technical committee, ISO/TC 268 (International Standards Organization n.d.), that focuses on sustainable cities and communities. This committee has developed a series of ISO standards to guide cities in their sustainability efforts. Under the direct responsibility of the ISO/TC 268 Secretariat, 16 standards have been published, and another 9 are under development. Additionally, 16 standards have been published under the subcommittee "Smart Community Infrastructures" and another 18 standards under the subcommittee "Sustainable Cities and Communities—Sustainable Mobility and Transportation." There are an additional 12 standards in total by the two subcommittees under development (International Standards Organization n.d.). Key groups within the ITU that are involved in the development of sustainable and smart cities include the ITU-T Study Group 20 (SG20) and the United for Smart Sustainable Cities (U4SSC) initiative. ITU-T Study Group 20 (SG20) focuses on the Internet of Things (IoT) and its applications in smart cities and communities (ITU-T n.d.). The U4SSC provides a platform for sharing best practices and experiences from cities around the world (United for Smart Sustainable Cities (U4SSC) n.d.). It helps cities implement ITU standards and KPIs, and publishes case studies and reports on the progress of cities in their smart city initiatives. In essence, SG20 provides the technical foundation for smart city technologies, while U4SSC focuses on the broader strategy and implementation aspects of achieving sustainable cities through technology.

Key standards developed by ISO/TC 268 for sustainable cities are the following:

- ISO 37100:2016, "Sustainable cities and communities—Vocabulary" (International Organization for Standardization 2016a).
- ISO 37101:2016, "Sustainable development in communities—Management system for sustainable development—Requirements with guidance for use" (International Organization for Standardization 2016b).
- ISO/TR 37121:2017, "Sustainable development in communities—Inventory of existing guidelines and approaches on sustainable development and resilience in cities" (International Organization for Standardization 2017a).
- ISO 37153:2017, "Smart community infrastructures—Maturity model for assessment and improvement" (International Organization for Standardization 2017b).
- ISO 37120:2018, "Sustainable cities and communities—Indicators for city services and quality of life" (International Organization for Standardization 2018).
- ISO 37122:2019, "Sustainable cities and communities—Indicators for smart cities" (International Organization for Standardization 2019a).
- ISO 37123:2019, "Sustainable cities and communities—Indicators for resilient cities" (International Organization for Standardization 2019b).
- ISO 37104:2019, "Sustainable cities and communities—Transforming our cities— Guidance for practical local implementation of ISO 37101" (International Organization for Standardization 2019c).
- ISO 37106:2021, "Sustainable cities and communities—Guidance on establishing smart city operating models for sustainable communities" (International Organization for Standardization 2021).

The international standard ISO 37101 (International Organization for Standardization 2016b) focuses on addressing the threats and achieving the 2030 targets for sustainable development set by the UN (SDGs). However, it does not provide sufficient guidance on strategy and the prioritization of goals for selecting improvement actions.

To support the sustainability efforts of Greek cities, the Hellenic Organization for Standardization developed the standard ELOT 1457, "Sustainable development in cities— Indicators for sustainability" (Hellenic Organization for Standardization 2018c). The Greek standard incorporates all the sustainability thematic sections and related indicators from the international standards. It follows very closely the structure and indicators of ISO 37120. Additionally, it includes themes for highlighting identity, leveraging comparative advantages, and Greek indicators.

Furthermore, ELOT is active in ISO and ITU, participating in working groups that develop new standards for sustainable cities. The ELOT 1457 standard and the organization's involvement in the development of new standards at ISO are discussed in the present work.

9.2 Background: ELOT and Sustainable Development in Cities

The Greek Standard ELOT 1457 was developed by the Technical Committee ELOT/TET 16/OE5, "Sustainable and Smart Cities" in collaboration with ELOT/TE 105/OE1, "ICT Indicators for Sustainable and Smart Cities". The ELOT 1457 standard proposes sustainability indicators in 15 themes across the four pillars of sustainability: Economy, Environment, Society, and Governance. To assist municipalities and city partners in choosing actions and setting priorities to achieve sustainable development goals, there is a mapping to the 17 UN goals.

The ELOT 1457 standard includes international indicators used for comparisons, statistics, and evaluations between cities at an international level. It also proposes indicators that meet the needs of Greek cities to improve their sustainability performance. The indicators can be used to establish the city's sustainability benchmark and monitor its performance progress after implementing transformation measures. To achieve sustainability, it is essential to consider all aspects of the city as a broader urban system, including impacts on peri-urban aquatic and terrestrial ecosystems. The design of transformation measures should account for current resource use and efficiency, as well as future resource needs and predicted disposal conditions, especially for non-sustainable or scarce resources.

The indicators have been developed to assist cities:

(a) to measure the performance management of their services and quality of life over time;
(b) to learn from each other by enabling comparison across a wide range of metrics and performance indicators;
(c) to share best practices;
(d) to enable comparisons, international statistics, and evaluation of cities.

For guidance on digital transformation in sustainable and smart cities, the ELOT 1457 standard includes Annex C with relevant indicators for Telecommunications, Information Technology, and Communications. These indicators are primarily based on standards from the International Telecommunication Union (ITU).

The ELOT 1457 adopts the principles and objectives of sustainability. It can be used in conjunction with ELOT ISO 37101:2018, "Sustainable cities—Management systems for sustainable cities—Requirements with guidance for use" (Hellenic Organization for Standardization 2018a), or other management system standards such as ELOT EN ISO 9001 for quality (Hellenic Organization for Standardization 2015a) or ELOT EN ISO 14001 (Hellenic Organization for Standardization 2015b) for environmental management.

The ELOT 1457 applies to any city, municipality, district, or other form of local government that aims to measure its performance in a comparable and verifiable manner, irrespective of size, management capacity, and geographical location.

9.2.1 Sustainability Themes

The four pillars of sustainable development are Economy, Environment, Society, and Governance. ELOT 1457 outlines 15 sustainability themes that fall under these four pillars.

Economy

>Theme 1. Economic development with social cohesion.
>Theme 2. Fiscal self-sufficiency.

Environment

>Theme 3. Environmental and energy issues. Theme 3 includes the following sub-themes:
>3.1 Greenhouse gas emissions, air quality, and noise.
>3.2 Protection of peri-urban ecosystems, in aquatic and terrestrial environments.
>3.3 Use of sustainable forms of energy and energy saving.
>Theme 4. Design of public spaces including public buildings, urban green spaces, and other open public areas e.g., recreational areas.
>Theme 5. Urban mobility or Transport/Transportation.
>Theme 6. Telecommunications, Information and Communication Technologies for smart infrastructures, networks, and services.
>Theme 7. Solid and liquid waste management and dissemination of circular economy principles.

Society

>Theme 8. Security and protection of citizens, public spaces, and dealing with emergencies.
>Theme 9. Housing and living conditions.
>Theme 10. Drinking water network and sanitary conditions.
>Theme 11. Health services.

Governance

>Theme 12. Open and participatory governance and equal opportunities.
>Theme 13. Openness of data, applications, and services.
>Theme 14. Knowledge and creativity. Education and innovation.
>Theme 15. Protection, promotion and exploitation of cultural heritage (for the attractiveness of a tourist destination).

9.2.2 Sustainability Performance Indicators

The ELOT 1457 has been designed to assist cities in guiding, measuring, and evaluating the sustainability, quality of services, and quality of life within urban areas. Performance indicators are completed and published annually. The indicators for each theme, where possible, are selected and combined based on inputs and outputs to achieve efficient resource management.

When interpreting the results of specific thematic sustainability assessments, it is important to consider various types of indicators that may be affected. Focusing on a single indicator can lead to misleading or incomplete conclusions. Expected outcomes should also be considered in the analysis of results.

Users can also consider additional parameters, which should be clearly defined and justified within the reporting document. These parameters may include indicators applicable to broader administrative areas like regions or metropolitan areas. As some indicators have an indirect connection to sustainability, it's essential to consider a city's resource use efficiency as well. When analyzing a city's holistic characteristics, indicators can be grouped thematically. Additionally, these indicators can be supplemented with other sets to achieve a more comprehensive and well-rounded approach to sustainability analysis.

For data interpretation purposes, cities should consider their operating context when interpreting the results. The institutional framework may affect the ability to manage certain indicators effectively. In some cases, services may be provided by government, private entities, or other community organizations.

The performance indicators provide the means for cities to self-assess. It is desirable for each city to continuously monitor the degree of achievement according to the targets set for the indicators. Depending on the value a performance indicator receives, conclusions can be drawn about the extent to which the UN goals it is associated with are being satisfied.

The presentation of several of the performance indicators follows, to provide the reader with a more comprehensive understanding of their purpose, how they are categorized based on the sustainability theme they belong to, and which UN Goals they are connected to. Initially, the 17 UN Sustainable Development Goals (SDGs) as presented in the "2030 Agenda for Sustainable Development" are indicated in Table 9.1 (Department of Economic and Social Affairs 2015). The "2030 Agenda for Sustainable Development" is a comprehensive plan of action adopted by all United Nations Member States in September 2015. It aims to address the global challenges we face, including poverty, inequality, climate change, environmental degradation, peace, and justice. The Agenda is built around 17 Sustainable Development Goals (SDGs) and 169 targets designed to guide global efforts toward sustainable development.

Table 9.2 presents the performance indicators that belong to Sustainability Theme 1, "Economic development with social cohesion", which is under the pillar of sustainable

Table 9.1 The 17 UN sustainable development goals (SDGs)

A/A	Goal
1	No poverty
2	Zero hunger
3	Good health and well-being
4	Quality education
5	Gender equality
6	Clean water and sanitation
7	Affordable and clean energy
8	Decent work and economic growth
9	Industry, innovation and infrastructure
10	Reduced inequalities
11	Sustainable cities and communities
12	Responsible consumption and production
13	Climate action
14	Life below water
15	Life on land
16	Peace, justice and strong institutions
17	Partnerships for the goals

development "Economy". These specific indicators are 14 in total and are connected to the UN Goals 1, 8, 9, 10, and 16.

Table 9.3 presents the performance indicators that belong to Sustainability Theme 3.1, "Greenhouse gas emissions, air quality, and noise", which is under the pillar of sustainable development "Environment". These specific indicators are 11 in total and are connected to the UN Goals 3, 12, 13, 14, and 15.

Table 9.4 presents the performance indicators that belong to Sustainability Theme 11, "Health", which is under the pillar of sustainable development "Society". These specific indicators are 17 in total and are connected to the UN Goals 3 and 12.

Table 9.5 presents the performance indicators that belong to Sustainability Theme 13, "Open data, applications, and services", which is under the pillar of sustainable development "Governance". These specific indicators are 20 in total and are connected to the UN Goals 5, 10, 16, and 17.

Table 9.2 Sustainability performance indicators—economy—Theme 1

Sustainability theme	Indicator	UN goals
Economy		
Theme 1. Economic development with social cohesion	1.1 Business activities per 100,000 inhabitants	1, 8, 9, 10 and 16
	1.2 City unemployment rate	
	1.3 Estimated value of commercial and industrial property as a percentage of total estimated property value	
	1.4 Percentage of the city's population living in poverty	
	1.5 Percentage of self-employed individuals	
	1.6 Percentage of employees on fixed-term contracts	
	1.7 Percentage of individuals in full-time employment	
	1.8 Youth unemployment rate	
	1.9 Number of visitors annually, per 100,000 population	
	1.10 Average household income	
	1.11 Percentage of the city's population using facilities such as social grocery stores, medical centers, or tutoring centers	
	1.12 Percentage of the city's population with disabilities living in poverty conditions	
	1.13 Number of business activities by people with disabilities	
	1.14 Annual population growth of the city	

9.3 Research Methodology: ELOT 1457—Indicators for Sustainability

9.3.1 ICT Indicators for Smart and Sustainable Cities

In ELOT 1457, in addition to the sustainability performance indicators, the Information and Communication Technology Key Performance Indicators (ICT KPIs) are also introduced. The purpose of establishing ICT KPIs is to use them as criteria for assessing the contribution of ICT to the creation of smarter and more sustainable cities and to provide the means for cities to self-assess. It is desirable for each city to continuously monitor the degree of achievement according to the targets set for the indicators.

ICT includes devices, networks, and services as well as ICT projects. At the city level, ICT projects have the specific purpose of spreading ICT to different parts of society

Table 9.3 Sustainability performance indicators—environment—Theme 3.1

Sustainability theme	Indicator	UN goals
Environment		
Theme 3.1 Greenhouse gas emissions, air quality, and noise	3.1.1 Concentration of fine particulate matter (PM2.5)	3, 12, 13, 14 and 15
	3.1.2 Concentration of particulate matter (PM10)	
	3.1.3 Greenhouse gas emissions measured in tons per capita	
	3.1.4 Concentration of NO_2 (nitrogen dioxide)	
	3.1.5 Concentration of SO_2 (sulfur dioxide)	
	3.1.6 Concentration of O_3 (ozone)	
	3.1.7 Noise pollution	
	3.1.8 Percentage of urban area covered by a digital noise monitoring application	
	3.1.9 Number of installations with digital noise monitoring application per km^2	
	3.1.10 Percentage of urban area covered by external particulate matter and toxic substance measurement systems	
	3.1.11 Number of external installations with ICT systems for air pollution monitoring per km^2	

to improve the city's sustainability. The indicators can be used to evaluate the city's sustainability before and after the implementation of ICT projects.

The indicators are distinguished into central (C), which apply to all cities, and additional (A), depending on the goals each city sets to improve its sustainability and intelligence. The goals can vary according to population growth, environmental conditions, demographics, etc.

The ICT KPIs have been selected from relevant documents published by the International and European Standardization Organizations (ISO, ITU, and the European Telecommunications Standards Institute (ETSI)). More specifically, from ISO: ISO 37120 (International Organization for Standardization 2018) and ISO 37122 (International Organization for Standardization 2019a). From ITU: Y.4901 (International Telecommunication Union 2016a), Y.4902 (International Telecommunication Union 2016b), and Y.4903 (International Telecommunication Union 2022). From ETSI: TS 103 463 (European Telecommunications Standards Institute 2017). The ICT KPIs are classified into one of the following themes:

Table 9.4 Sustainability performance indicators—society—Theme 11

Sustainability theme	Indicator	UN goals
Society		
Theme 11: Health	11.1 Average life expectancy	3 and 12
	11.2 Hospital beds per 100,000 inhabitants	
	11.3 Doctors per 100,000 inhabitants	
	11.4 Under-five mortality rate per 1000 births	
	11.5 Percentage of regular smokers relative to the city's population	
	11.6 Percentage of city residents who experienced psychological distress in the past few months	
	11.7 Percentage of residents who reported having a traffic accident, resulting in an injury requiring medical attention, in the past 12 months	
	11.8 Percentage of traffic accidents in the city involving alcohol/drugs	
	11.9 Nursing and midwifery staff per 100,000 inhabitants	
	11.10 Number of medical and nursing staff	
	11.11 Suicide rate per 100,000 inhabitants	
	11.12 Percentage of young adults aged 18–24 with a body mass index (BMI) of 30 or higher	
	11.13 Percentage of pupils who reported being under the influence of alcohol	
	11.14 Percentage of city residents with electronic health records (EHRs)	
	11.15 Percentage of city residents with electronic medical records (EMRs)	
	11.16 Percentage of hospitals, pharmacies, and healthcare service providers using ICT for sharing medical resources such as hospital beds and medical information, specifically electronic medical records	
	11.17 Percentage of patients involved in telemedicine programs, including services such as online consultations, monitoring, online medical advice, and healthcare guidance, etc.	

Theme 1. Information and Communication Technology (ICT).
Theme 2. Utilization and promotion of cultural heritage.
Theme 3. Health.
Theme 4. Innovation.
Theme 5. Education.

Table 9.5 Sustainability performance indicators—governance – Theme 13

Sustainability theme	Indicator	UN goals
Governance		
Theme 13: Openness of data, applications, and services	13.1 Percentage of accessible public services	5, 10, 16 and 17
	13.2 Percentage of accessible public recreation spaces	
	13.3 Percentage of businesses with interoperable open databases	
	13.4 Percentage of private recreational/cultural spaces	
	13.5 Percentage of accessible educational facilities	
	13.6 Percentage of accessible healthcare facilities	
	13.7 Percentage of employees with disabilities in the private sector	
	13.8 Percentage of accessible temporary accommodation tourist infrastructure	
	13.9 Annual number of tourists with disabilities	
	13.10 Number of organizations providing open data per institutionalized thematic area as defined by the European Union	
	13.11 Number of online databases of public libraries per 100,000 population	
	13.12 Number of new digital applications for multiple devices (desktop, mobile, etc.) provided per municipality and per year	
	13.13 Number of new digital applications annually for multiple devices (desktop or mobile, etc.) in the municipality	
	13.14 Percentage of urban information available online and the existence of ICT systems for easy access and anonymous feedback mechanisms that empower cities to improve their governance	
	13.15 Percentage of city residents using online information and the percentage using feedback mechanisms with ICT	
	13.16 Percentage of city residents using online public services and facilities (e.g., school selection, booking public sports facilities, library services, etc.)	
	13.17 Percentage of available open data from cities	
	13.18 Number of open government datasets per 100,000 residents	

(continued)

Table 9.5 (continued)

Sustainability theme	Indicator	UN goals
	13.19 Existence of a framework enabling the use of public city-data	
	The extent of disclosure of administrative information	

Theme 6. Openness of data, applications, and services.
Theme 7. Energy.
Theme 8. Electronic services.
Theme 9. Information and Communications Technology (ICT) security.
Theme 10. Electromagnetic fields.
Theme 11. Capital investments.
Theme 12. Public safety and emergencies.
Theme 13. Environmental indicators.
Theme 14. Noise management.
Theme 15. Wastewater.
Theme 16. Greenhouse gas emissions and air quality.
Theme 17. Road infrastructure and transportation.
Theme 18. Building management.

The presentation of several ICT KPIs follows, aiming to provide the reader with a more comprehensive understanding of their purpose, how they are categorized based on their respective themes, whether they are classified as central (C) or additional (A), and their alignment with specific targets within the United Nations Sustainable Development Goals (SDGs).

Table 9.6 presents the ICT KPIs that belong to Theme 1, "ICT", which total 19 in number.

Table 9.7 presents the ICT KPIs that belong to Theme 4, "Innovation", which total 8 in number.

Table 9.8 presents the ICT KPIs that belong to Theme 7, "Energy", which total 5 in number.

Table 9.9 presents the ICT KPIs that belong to Theme 13, "Environmental indicators", which total 8 in number.

Table 9.10 presents the ICT KPIs that belong to Theme 17, "Road infrastructure and transportation", which total 7 in number.

Table 9.11 presents the ICT KPIs that belong to Theme 18, "Building management", which total 2 in number.

Table 9.6 ICT KPIs—Theme 1: ICT

Theme	Indicator	Description	Reference	C	A	UN goal
ICT	Availability of computers or similar devices	Percentage of households with at least one computer or similar device (tablet, smartphones, etc.)	Y.4901/ I1.1.1 Y.4903/ C1.1.2	x		9.c
	Availability of smartphones and tablets	Number of smart phones and tablets per 100 inhabitants	Y.4901/ A1.1.8		x	
	Availability of mobile-cellular telephones	Mobile-cellular telephone subscriptions per 100 inhabitants	Y.4901/ A1.1.1		x	
	Households with a mobile device	Percentage of households with at least one smartphone or similar device	Y.4903/ A1.1.3 ISO 37120		x	9.c
	Availability of internet access in households	Percentage of households with internet access for any household member via a fixed or mobile network at any given time	Y.4901/ I1.1.2 Y.4903/ C1.1.1 ISO 37120	x		9.c 17.8
	Availability of fixed broadband subscriptions	Fixed (wired) broadband subscriptions per 100 inhabitants	Y.4901/ I1.1.3 Y.4903/ A1.1.2 TS 103 463	x		9.c
	Availability of ultra high speed wireline connection	Percentage of households with access to downstream speeds equal to, or greater than, 30 Mbits/s	Y.4901/ A1.1.5		x	
	Quality of fixed broadband	Mean-download speed (fixed)	Y.4901/ A1.1.9		x	
	Availability of wireless broadband subscriptions	Wireless broadband subscriptions per 100 inhabitants	Y.4901/ I1.1.4 Y.4903/ A.1.1.1	x	x	9.c 5.b
	Quality of mobile broadband	Cell-edge performance (mobile)	Y.4901/ A1.1.10		x	

(continued)

Table 9.6 (continued)

Theme	Indicator	Description	Reference	C	A	UN goal
	Availability of high-speed mobile broadband	Percentage of city area which provides access to downstream speeds equal to, or greater than, 10 Mbits/s	Y.4901/ A1.1.6		x	
	International internet bandwidth	International internet bandwidth (bit/s) per internet user	Y.4901/ A1.1.2		x	
	Use of internet by city inhabitants	Percentage of inhabitants using internet	Y.4901/ A1.1.3		x	
	Coverage rate of digital broadcasting network	Percentage of digital broadcasting network covering families in the city	Y.4901/ A1.1.4		x	
	Availability of WiFi in public areas	Number of WiFi hotspots at certain points in the city centre	Y.4901/ A1.1.7		x	
	Access to public free WiFi	WiFi coverage of public spaces	TS 103 463			
	Household expenditure on ICT	Percentage of household expenditure on ICT	Y.4902/ I3.7.1	x		
	Total urban area served by publicly available internet access	Percentage of urban area with publicly available internet access	ISO 37122			
	Quality of publicly available internet connection	Average broadband speed				

9.3.2 ELOT's Role in Developing and Promoting International Standards for Sustainable Development in Cities

The Hellenic Organization for Standardization (ELOT) is responsible for developing, publishing, and promoting national standards in Greece, aligning with International and European standards to ensure quality, safety, and efficiency across various sectors. ELOT actively participates in international working groups within the ISO and ITU that prepare standards for sustainable development in cities and communities. Through its involvement, ELOT contributes to the development of global standards by sharing expertise and collaborating with other member countries. This participation ensures that Greek perspectives and needs are represented, while also allowing Greece to stay at the forefront of adopting and implementing cutting-edge standards in areas such as sustainable development.

Table 9.7 ICT KPIs—Theme 4: innovation

Theme	Indicator	Description	Reference	C	A	UN goal
Innovation	Investments in ICT innovation	Percentage of private sector spending on ICT innovation investments	Y.4902/I3.8.2	x		
	Research and development intensity in ICT	Percentage of research and development intensive ICT companies among all companies	Y.4901/I3.8.1	x		
	Intangible investments as a percentage of GDP	Percentage of intangible investments (e.g., research and development, software, design, marketing, education and training) in new and existing businesses expressed as percentage of city GDP	Y.4901/I3.9.1	x		
	Intangible investments in comparison with total investments	Percentage of intangible investments (e.g., research and development, software, design, marketing, education and training) in new and existing businesses related to overall investments	Y.4901/A3.9.1		x	
	ICT-related patents	Number of ICT-related patents granted per capita	Y.4902/I3.8.3	x		
	Innovation hubs in the city	Number of innovation hubs in the city, whether private or public, per 100,000 inhabitants	TS 103 463			

(continued)

Table 9.7 (continued)

Theme	Indicator	Description	Reference	C	A	UN goal
	Employees belonging to ICT sector	Percentage of employees in ICT sector among all employees	Y.4901/I3.9.2	x		
	Small and medium-sized enterprises	Percentage of small and medium-sized enterprises	Y.4903/A1.2.1		x	9.3 8.3

Table 9.8 ICT KPIs—Theme 7: energy

Theme	Indicator	Description	Reference	C	A	UN goal
Energy	Availability of smart electricity meters	Percentage of the electricity consumers (including households, companies, etc.) with ICT based electricity meters	Y.4901/ I6.3.1	x		
	Electricity supply system management using ICT	Percentage of power substation and user points under automatic inspection using ICT	Y.4901/ A6.3.1 Y.4903/ A1.6.3		x	
	Availability of visualized real-time information regarding electricity use	Percentage of users with real-time information on quantum of electricity usage and electricity use pattern	Y.4901/ A6.3.2		x	
	Gas system management using ICT	Percentage of gas supply systems under automatic monitoring using ICT	Y.4901/ I6.8.5	x		
	Availability of visualized real-time information regarding gas use	Percentage of users with real-time information on quantum of gas usage and gas use pattern	Y.4901/ A6.8.1		x	

9.3.3 ELOT's Active Role in Shaping ISO Standards for Sustainable Urban Development and Disaster Risk Reduction

ELOT participates in the preparation of standards within ISO. Specifically, it has teams of carefully selected scientists who participate in working groups for the production of standards. ELOT participates in ISO as a Participating Member (P-member), which entails certain privileges as well as obligations. Privileges include voting rights on the technical

Table 9.9 ICT KPIs—Theme 13: environmental indicators

Theme	Indicator	Description	Reference	C	A	UN goal
Environmental indicators	Water supply system management using ICT	Percentage of the water supply systems under automatic monitoring using ICT so as to ensure water quality and reduce leakage	Y.4901/I6.1.1	x		
	City freshwater sources monitored using ICT	Percentage of the city's fresh water sources monitored using ICT with respect to availability	Y.4901/I6.1.2	x		
	Application of city water monitoring through ICT	Percentage of the city water resources (rivers, lakes etc.) monitored by ICT with respect to water pollution and quality	Y.4901/I2.5.1	x		
	Monitoring of water supply with ICT	Percentage of the water distribution system monitored by ICT	Y.4903/A1.6.2		x	
	Availability of smart water meters	Percentage of the water consumers (including households, companies, etc.) with ICT based water meters	Y.4901/I6.1.3 Y.4903/C1.6.1	x		9.1
	Availability of visualized real-time information regarding water use	Percentage of users with real-time information on quantum of water usage and water use pattern	Y.4901/A6.1.1		x	
	Drainage system management using ICT	Percentage of the drainage systems monitored in real-time using ICT	Y.4901/I6.2.2	x		
	Drainage system management	Percentage of the drainage systems monitored using ICT	Y.4903/A2.2.2		x	6.5 6.4

content of standards, drafts, and other documents produced by committees, and obligations include actively contributing to the development of ISO standards by attending meetings, commenting on documents, and providing technical expertise.

Table 9.10 Theme 17: road infrastructure and transportation

Theme	Indicator	Description	Reference	C	A	UN goal
Road infrastructure and transportation	Availability of traffic monitoring using ICT	Percentage of streets with traffic monitoring using ICT (e.g., using sensors to produce traffic volume maps, etc.)	Y.4901/I6.8.1	x		
	Traffic monitoring	Percentage of main roads monitored by ICT	Y.4903/A1.6.6		x	9.1
	Availability of parking guidance systems	Percentage of parking lots and street parking spaces with ICT based parking guidance systems	Y.4901/I6.8.2	x		
	Street lighting management using ICT	Percentage of street lamps under automatic management using ICT (e.g., light/sound control and solar power charging)	Y.4901/I6.8.4	x		
	Availability of online bike/car sharing system	Percentage of city area covered by an online bike/car sharing system	Y.4901/A6.8.2		x	
	Use of real-time navigation	Percentage of real-time navigation users compared to all navigation system users	Y.4901/A6.8.3		x	
	Real-time public transport information	Percentage of public transport stops and stations providing real-time traffic information	Y.4903/C1.6.7	x		11.2

Table 9.11 Theme 18: building management

Theme	Indicator	Description	Reference	C	A	UN goal
Building management	Automatic energy management in buildings	Percentage of public and private sector buildings using ICT based systems to automatically regulate and reduce energy needs	Y.4901/I6.11.1	x		
	Integrated management in public buildings	Proportion of public buildings using integrated ICT systems to automate building management and create flexible, effective, comfortable and secure environment	Y.4901/I6.11.2	x		

Regarding the production of standards related to the sustainable development of cities and communities, which is the focus of ISO/TC 268 (Technical Committee), ELOT participates in Working Group 6 (WG 6) under Subcommittee 1 (SC 1), which deals with the production of standards concerning disaster risk reduction. The full code and title of the working group are ISO/TC 268/SC 1/WG 6, "Disaster risk reduction" (International Standards Organization n.d.). ELOT's participation in this working group began in June 2021, and it has actively contributed to the development and production of two standards: ISO/TR 6030:2022, "Smart community infrastructures—Disaster risk reduction—Survey results and gap analysis" (International Organization for Standardization 2022), and ISO 37174:2024, "Smart community infrastructures—Disaster risk reduction—Guidance for implementing seismometer systems" (International Organization for Standardization 2024). ELOT's participation in the production of these standards was active, involving attending meetings, commenting on documents, providing technical expertise, and balloting.

9.4 Conclusions

Modern cities face numerous challenges that threaten the well-being of their inhabitants. While international organizations like ISO and ITU provide frameworks for sustainable urban development (e.g., ISO 37100 series), successful implementation requires localized strategies. Recognizing this, the Hellenic Organization for Standardization (ELOT) has played a crucial role in supporting Greek cities by offering localized versions of international standards (ELOT ISO 37100 and ELOT ISO 37101) and developing the ELOT 1457 standard.

The ELOT 1457 standard represents a robust framework developed by ELOT to guide Greek cities towards sustainable development. Divided into four thematic pillars—Economy, Environment, Society, and Governance—the standard employs comprehensive sustainability performance indicators aligned with the UN Sustainable Development Goals (SDGs). These indicators enable cities to measure and enhance their performance across various dimensions, from economic resilience and environmental quality to public health and governance transparency. This holistic approach ensures that cities do not focus solely on economic growth but also consider social equity, environmental quality, and effective governance as integral to sustainable development. By aligning with the UN Sustainable Development Goals (SDGs), these indicators provide a recognized framework for cities to benchmark their progress. This alignment helps cities contribute to global efforts while addressing local challenges and priorities. Indicators enable evidence-based decision-making by providing cities with quantifiable data on their performance across different dimensions. This helps policymakers prioritize interventions and allocate resources effectively to achieve sustainability targets. The emphasis on continuous monitoring and evaluation encourages cities to set ambitious yet achievable targets and regularly assess their progress. This iterative process supports cities in adapting to changing circumstances and emerging challenges.

In addition to sustainability performance indicators, ELOT 1457 incorporates Information and Communication Technology Key Performance Indicators (ICT KPIs). These KPIs are designed to evaluate the impact of ICT initiatives on city sustainability, promoting smarter urban environments. Categorized into multiple themes such as ICT, Innovation, Energy, Transport infrastructure, and Building management, the ICT KPIs facilitate continuous monitoring and improvement in areas crucial for urban development. Incorporating ICT KPIs alongside traditional sustainability indicators enhances cities' capacity for innovation and efficiency. ICT tools and data analytics enable real-time monitoring, predictive modeling, and decision support systems, thereby optimizing resource use and enhancing service delivery. ICT KPIs align city strategies with global agendas such as the UN SDGs and regional frameworks. They provide a structured approach for cities to measure how ICT investments contribute to economic development, environmental sustainability, social equity, and effective governance. By measuring ICT infrastructure availability, connectivity, and digital service accessibility (e.g., broadband penetration, mobile device usage), cities can enhance service delivery across various sectors. This includes healthcare, education, transportation, and public safety, leading to improved quality of life for residents. ICT KPIs also encompass environmental indicators such as energy efficiency (e.g., smart metering, energy management systems) and environmental monitoring (e.g., air and water quality sensors). By integrating sustainability into ICT initiatives, cities can mitigate environmental impacts and promote resource efficiency. ICT KPIs related to infrastructure management (e.g., smart grids, traffic monitoring systems, smart buildings) can optimize resource utilization and operational efficiency. These technologies enable cities to manage infrastructure assets more sustainably and respond effectively to urban challenges.

Overall, ELOT 1457 not only provides a structured approach for Greek cities to assess their sustainability but also integrates ICT advancements to foster innovation and efficiency. Through adherence to these standards, Greece aims to achieve resilient, inclusive, and sustainable urban landscapes that meet the challenges of the twenty-first century.

Furthermore, ELOT's active participation in international standardization efforts, particularly within ISO Technical Committee 268, underscores Greece's commitment to global sustainability practices. By contributing to standards like ISO/TR 6030 and ISO 37174, which focus on disaster risk reduction and smart community infrastructures, ELOT ensures that Greek cities benefit from cutting-edge methodologies and best practices in sustainable urban development.

9.5 Revision Questions with Answers

1. How does the UN Sustainable Development Goal 11 define sustainable cities?
According to UN Sustainable Development Goal 11, sustainable cities are those dedicated to achieving green sustainability, social sustainability, and economic sustainability.

2. How are smart cities and sustainable cities interconnected?
Sustainable cities and smart cities are interconnected through their shared goals of improving urban quality of life while minimizing environmental impact. Sustainable cities focus on resource efficiency and social equity through strategies like renewable energy and green infrastructure. Smart cities use technology and data to optimize urban operations, enhance service delivery, and promote sustainability. Both aim to create resilient, livable cities that meet the needs of current and future generations effectively.

3. Name three of the key standards developed by ISO/TC 268 for sustainable cities and communities.

- ISO 37100:2016, "Sustainable cities and communities—Vocabulary"
- ISO 37101:2016, "Sustainable development in communities—Management system for sustainable development—Requirements with guidance for use"
- ISO 37120:2018, "Sustainable cities and communities—Indicators for city services and quality of life".

4. What are the primary objectives and benefits of implementing the ELOT 1457 standard for sustainable and smart cities?

The ELOT 1457 standard aims to provide sustainability indicators across 15 themes aligned with the pillars of Economy, Environment, Society, and Governance. It assists cities in benchmarking their sustainability performance, monitoring progress, and fostering international comparisons and best practices sharing. The standard supports cities in setting priorities, implementing transformation measures, and ensuring sustainable urban development.

5. What is the purpose of establishing information and communication technology key performance indicators (ICT KPIs) in ELOT 1457, and how are they intended to be used by cities?

The purpose of establishing ICT KPIs is to use them as criteria for assessing the contribution of ICT to the creation of smarter and more sustainable cities and to provide the means for cities to self-assess. It is desirable for each city to continuously monitor the degree of achievement according to the targets set for the indicators. The indicators can be used to evaluate the city's sustainability before and after the implementation of ICT projects.

6. Which specific standards has ELOT contributed to under the ISO/TC 268/SC 1/WG 6, and what was the nature of its contributions?

ELOT has contributed to the development and production of two specific standards under the ISO/TC 268/SC 1/WG 6. These standards are ISO/TR 6030:2022, "Smart community infrastructures—Disaster risk reduction—Survey results and gap analysis," and ISO 37174:2024, "Smart community infrastructures—Disaster risk reduction—Guidance for implementing seismometer systems". ELOT's contributions involved attending meetings, commenting on documents, providing technical expertise, and participating in balloting processes.

References

Anthopoulos LG (2017) Understanding smart cities: a tool for smart government or an industrial trick?, vol. 22, p. 293. Springer International Publishing, Cham, Switzerland. https://link.springer.com/content/pdf/10.1007/978-3-319-57015-0.pdf

Department of Economic and Social Affairs (2015) Sustainable development goals. United Nations. https://sdgs.un.org/

European Telecommunications Standards Institute (2017) TS 103 463, access, terminals, transmission and multiplexing (ATTM); Key performance indicators for sustainable digital multiservice cities. https://www.etsi.org/deliver/etsi_ts/103400_103499/103463/01.01.01_60/ts_103463v010101p.pdf

Hellenic Organization for Standardization (2015a) ELOT EN ISO 9001 E4, Quality management systems—requirements. http://sales.elot.gr/online/search/details.do?documentId=300010000050894

Hellenic Organization for Standardization (2015b) ELOT EN ISO 14001 E3, Environmental management systems—requirements with guidance for use. http://sales.elot.gr/online/search/details.do?documentId=300010000050899

Hellenic Organization for Standardization (2018a) ELOT ISO 37101: Sustainable cities—management systems for sustainable cities—requirements with guidance. http://193.218.125.20/elot_iso_37101-el_2017_(2017_08_29).pdf

Hellenic Organization for Standardization (2018b) ELOT ISO 37100: Sustainable cities—terms and definitions of concepts. https://eshop.elot.gr/product/51446

Hellenic Organization for Standardization (2018c) ELOT 1457: Sustainable development in cities—indicators for sustainability. http://193.218.125.20/ELOT1457_20180712_ENQ.pdf

International Organization for Standardization (2014) ISO/IEC JTC 1, information technology. Smart cities: preliminary report 2014. http://www.iso.org/iso/smart_cities_report-jtc1.pdf

International Organization for Standardization (2016) ISO 37100:2016: Sustainable cities and communitie—Vocabulary. https://www.iso.org/standard/71914.html

International Organization for Standardization (2016) ISO 37101:2016, Sustainable development in communities—management system for sustainable development—requirements with guidance for use. https://www.iso.org/standard/61885.html

International Organization for Standardization (2017a) ISO/TR 37121:2017, Sustainable development in communities—inventory of existing guidelines and approaches on sustainable development and resilience in cities. https://www.iso.org/standard/63790.html

International Organization for Standardization (2017b) ISO 37153:2017, Smart community infrastructures—maturity model for assessment and improvement. https://www.iso.org/standard/69225.html

International Organization for Standardization (2018) ISO 37120:2018, Sustainable cities and communities—indicators for city services and quality of life. https://www.iso.org/standard/68498.html

International Organization for Standardization (2019a) ISO 37122:2019, Sustainable cities and communities—indicators for smart cities. https://www.iso.org/standard/69050.html

International Organization for Standardization (2019b) ISO 37123:2019, Sustainable cities and communities—indicators for resilient cities. https://www.iso.org/standard/70428.html

International Organization for Standardization (2019c) ISO 37104:2019, Sustainable cities and communities—transforming our cities—guidance for practical local implementation of ISO 37101. https://www.iso.org/standard/69895.html

International Organization for Standardization (2021) ISO 37106:2021, Sustainable cities and communities—guidance on establishing smart city operating models for sustainable communities. https://www.iso.org/standard/82854.html

International Organization for Standardization (2022) ISO/TR 6030:2022, Smart community infrastructures—disaster risk reduction—survey results and gap analysis. https://www.iso.org/standard/81941.html

International Organization for Standardization (2024) ISO 37174:2024, Smart community infrastructures—disaster risk reduction—guidance for implementing seismometer systems. https://www.iso.org/standard/69260.html

International Organization for Standardization. ISO 37100 series: Sustainable cities and communities. ISO/TC 268. https://www.iso.org/committee/656906/x/catalogue/p/1/u/0/w/0/d/0#projects

International Standards Organization (n.d.) Working groups by ISO/TC 268/SC 1. Smart community infrastructures. https://www.iso.org/committee/656906.html

International Standards Organization (n.d.) ISO/TC 268, Sustainable cities and communities. International Organization for Standardization. https://www.iso.org/committee/656906.html

International Telecommunications Union (2014) Setting the framework for an ICT architecture of a smart sustainable city. http://www.itu.int/en/ITU-T/focusgroups/ssc/Documents/website/web-fg-ssc-0345-r5-ssc_architecture.docx

International Telecommunication Union (2016a) ITU-T Y.4901, Key performance indicators related to the use of information and communication technology in smart sustainable cities. https://www.itu.int/itu-t/recommendations/rec.aspx?rec=12661

International Telecommunication Union (2016b) ITU-T Y.4902, Key performance indicators related to the sustainability impacts of information and communication technology in smart sustainable cities. https://www.itu.int/itu-t/recommendations/rec.aspx?rec=12662

International Telecommunication Union (2022) ITU-T Y.4903, Key performance indicators for smart sustainable cities to assess the achievement of sustainable development goals. https://www.itu.int/ITU-T/recommendations/rec.aspx?rec=14173&lang=en

International Telecommunication Union Telecommunication Standardization Sector (ITU-T) (n.d.) ITU-T study group 20: internet of things (IoT) and smart cities and communities (SC&C). https://www.itu.int/en/ITU-T/about/groups/Pages/sg20.aspx

United for Smart Sustainable Cities (U4SSC) (n.d.) About U4SSC. https://u4ssc.itu.int/

United Nations (2015) Goal 11: Sustainable cities and communities. https://sdgs.un.org/goals/goal11

Part V
Cases of Defining or Applying Smart City Standards

Brazilian Smart City Standards—Challenges and Opportunities in the Adaptation and Expansion of the SSCMM-ITU: Platform inteli.gente Management and Governance System for Digital Transformation and Sustainable Development

Luísa Paseto, Márcia Regina Martins Martinez, Ricarda Carolina Rende, Rodrigo Barbosa Paula, and Andre Carlos Ponce de Leon Ferreira de Carvalho

Abstract

Since 2020, the Ministry of Science, Technology and Innovation has coordinated of developing and updating the inteli.gente platform, designed to diagnose, expand, and implement indicators related to smart cities for the assessment of 5570 Brazilian cities, customized for each one of them. Online and free, the inteli.gente platform is a management and governance system with diagnoses and recommendations on the level

---

L. Paseto (✉) · M. R. M. Martinez · A. C. P. de Leon Ferreira de Carvalho
Artificial Intelligence Recreating Environments—IARA, University of São Paulo-USP, São Carlos, SP, Brazil
e-mail: lu.paseto@usp.br

M. R. M. Martinez
e-mail: marcia.martinez@usp.br

A. C. P. de Leon Ferreira de Carvalho
e-mail: andre@icmc.usp.br

R. C. Rende
National Institute of Telecommunications—INATEL, Minas Gerais, MG, Brazil
e-mail: ricarda@anatel.gov.br

R. B. Paula
Institute of Higher Education FUCAPI, Manaus, AM, Brazil
e-mail: rodrigodepaula@anatel.gov.br

of maturity in dimensions of sustainable development and institutional capacities. An adaptation and extension of the Smart Sustainable City Maturity Model of the International Telecommunication Union, it uses data science techniques to analyze the indicators.

Keywords

Smart cities • Maturity • Sustainability • Governance • Management

10.1 Introduction

The planning and actions to make a smart city are increasingly present on the agendas of sustainable development and digital transformation around the world. The concepts of smart cities emerge from the development of information and communication technologies (ICT) and the need for their incorporation into public services governance, both in urban and rural environments. According to Bibri and Krogstie (2017), smart city models have been developed for cities and countries through standardization and assessments for defining their maturity level and even for comparison (ranking) proposals.

The Ministry of Science, Technology, and Innovation (MCTI) of Brazil, concerned with seeking intelligent and sustainable solutions for its cities, chose to adapt and expand the Smart Sustainable Cities Maturity Model (SSCMM-ITU), developed by the International Telecommunication Union (ITU). At that time, it was understood that the SSCMM-ITU was flexible and applicable in different contexts, allowing them to make decisions regarding the use of financial and infrastructure cities resources.

Therefore, since 2020, MCTI has been guiding the inteli.gente platform construction, with a focus on diagnosing, expanding, and implementing indicators for smart cities in Brazil, customized for each of its 5570 municipalities.

The inteli.gente platform is an online and free system available for all Brazilian cities. The platform's diagnoses and recommendations are divided into three dimensions of sustainable development: economy, environment, and socio-cultural, as well as an institutional capacity dimension. These four dimensions can be used for public policy management and governance.

The status of Brazil as a member state of ITU has allowed this experience to be shared worldwide through the development of a document with the Brazilian use case in the implementation of SSCMM-ITU.

The first version of inteli.gente was developed between January 2020 and October 2021 (CTI/poli.TIC 2020). However, IARA-USP one of the eight National Applied Artificial Intelligence Research Centers supported by MCTI, has been transforming the inteli.gente platform into a Data Lake for sustainable development and digital transformation since November 2021. In this way, it has been possible to operationalize the considerations and contributions obtained from participating Brazilian cities between April 2021 and February 2023.

Through IARA, it was possible to map out and improve institutional public policies using data provided by cities. Collaboratively, it also promoted cooperation, exchange, coordination and the creation of innovative initiatives related to digital transformation and sustainable development in the public sector.

The smart cities implementation in Brazil was carried out through the analysis of SSCMM—ITU and references gathered from experts, academics, and policymakers during the initial years (from 2020 to 2021). It was realized that there was a need to expand and adapt any of the maturity levels and the set of indicators offered by SSCMM-ITU.

The methodology used in the expansion and adaptation of SSCMM-ITU followed the studies reported by Yigitcanlar et al. 2019), which consider that public participation in decision-making and collaboration among all stakeholders ensure progress toward sustainability.

Thus, two new maturity levels and a new dimension, called institutional capability, were added. In the selection of indicators, over 1000 indicators from various sources were analyzed, with the most significant ones being: Cities in Motion Index (IESE), Program for Smart and Sustainable Cities (PCSI), Sustainable Development Goals (SDGs), ISO 37120, ISO 37122, ISO 37123; SEADE Foundation; CNM; CETIC br. From this evaluation, initially, 134 indicators were selected, and currently, 113 indicators have been selected. Due to its vast territorial expanse, Brazil presents high regional and local differences. While laws stipulate that access to information and electronic services should be available on municipal websites, there are still varying realities among cities. In this diverse context, the inteli.gente platform is considered strategic as it enables the understanding of these differences and realities, guiding the process of planning and constructing effective and efficient public policies for Brazilian cities.

10.2 Background

The inteli.gente platform follows the ITU-T recommendations: Smart Sustainable Cities Maturity Model (Y.4904) and Assessment framework for digital transformation of sectors in smart cities (Y.4906). It provides public administrators with recommendations and guidelines for building a sustainable city, along with their own diagnosis, strategies, indicators, and evolution goals. Furthermore, it has enabled the creation of a unique public policy database, considering infrastructure, intelligence, and sustainability for all Brazilian cities.

To achieve this, inteli.gente platform utilizes 113 indicators divided into four dimensions. The first three dimensions, economic, sociocultural, and environment, are related to sustainability, while the fourth dimension pertains to the institutional capacities of municipal public management. Additionally, the platform provides sociodemographic information, digital transformation insights, and institutional data on municipal management and governance, referred to as characterization.

As in the SSCMM-ITU, the indicators measurement on the inteli.gente platform adopts score levels based on the performance goals achievement, defined based on the Brazilian average performance, considering various relevant thematic areas of public policies for cities.

The maturity levels provide a diagnosis for cities that seek their own development in the short, medium, and long term. Each level has its objectives, indicators, and expected practices organized by dimensions and topics, presenting an evolutionary diagnosis for sustainable smart cities.

The maturity levels provide a diagnosis that can be used by the cities to plan their development in the short, medium, and long term. Each level has its objectives, indicators, and expected practices organized by dimensions and topics, providing a diagnosis of how sustainable and smart each city is along the time.

The first modification in the Brazilian model was the addition of two maturity levels in the lower layer of the SSCMM-ITU. The remaining SSCMM-ITU maturity levels known as levels, planning, alignment, development, integration, and optimization, were kept similar. The two new maturity levels were called "adhesion" and "commitment" with the aim of encompassing the less advanced Brazilian cities in terms of sustainable development and digital transformation. Therefore, the creation of these two additional levels was necessary due to the regional and local disparities among Brazilian cities, allowing the participation of all Brazilian cities in the developed model.

10.3 Research Methodology: Case Study

10.3.1 Strategic Level

The inteli.gente platform takes into account the diversities of Brazilian municipalities, through a digital tool that is open and scalable. In order to provide diagnostics and recommendations, the inteli.gente platform allows for prescriptive and descriptive, analyses, fostering the exchange of experiences and knowledge with other countries in similar contexts.

The SSCMM-ITU customization to the Brazilian reality was grounded in qualitative analysis and methodologies for evaluating public policies, which encompassed data collection, processing, analysis, and validation. It was based on an understanding of the local context through a multidimensional and interrelated perspective, incorporating the collection of various types of data and information (Lejano 2012).

A research study was conducted to understand the Brazilian diversity, employing qualitative and quantitative methods through primary and secondary data collection. A questionnaire was developed based on the data obtained from technical visits for primary data extraction. The information was analyzed using thematic, content, and statistical techniques. One of the findings revealed that it would not be feasible to include all Brazilian municipalities in the assessment of smart cities using only five maturity levels.

To use the SSCMM-ITU, a city must have a strategic plan for sustainable smart cities. However, during the technical visits, it was observed that many municipal managers do not formalize their actions through a planning process. Therefore, two additional levels of maturity were incorporated into the original SSCMM-ITU. The first added level is called "adhesion" which includes cities models that intend to adopt the Sustainable smart City but are still becoming familiar with the subject. The second added level called "commitment" which includes municipalities that already have knowledge about the subject and are seeking to develop strategies and plans for a smart city transformation.

The inteli.gente platform indicators construction involved gathering recommendations through meetings among academic experts, city managers, and e-government officials.

From this perspective, the selection of indicators for inteli.gente followed seven criteria:

- Alignment with the Brazilian cities necessities and consideration of the ITU reference framework, which was constructed based on countries with different realities;
- Compliance with the requirements of an indicator for sustainable development;
- Availability, whenever possible, in secondary databases from official Brazilian agencies to ensure reliability, validation, and periodic updates;
- Effectiveness in planning actions and strategies for the construction of public policies for a smart city;
- Incorporation of city diagnostics in terms of services and their expansion using ICT;
- Adherence to the Brazilian Charter for Smart Cities (CBCI 2020), specifically to objective 8.3: "Develop and provide a Brazilian Maturity System for Smart on a dedicated digital platform to be created and maintained by the federal government."

The aim of these criteria was to ensure that the selected indicators were suitable for the Brazilian context and supported the development of sustainable and smart cities.

By using suitable methodologies and indicators tailored to the specific characteristics of Brazilian cities and the municipal typologies of the National Urban Development Policy (PNDU), the system can effectively assist municipalities in their efforts to promote sustainable urban development and digital transformation. Furthermore, it enables national monitoring of the progress made in these areas, allowing for a comprehensive assessment of the impact and effectiveness of the implemented actions.

Simultaneously, the indicators selection aimed to incorporate concepts, features, and analytical frameworks that would include all 5570 Brazilian municipalities in diagnosing their maturity levels. Aspects of reliability, validity, sensitivity, comprehensiveness, and equity were also considered during these indicators construction (IISD-Bellagio 1997).

The maturity level diagnostics through the inteli.gente platform aims to provide a systemic results analysis by understanding the cross-relationships among indicators from various thematic areas (topics) of the public policy (Duncan et al. 1998; Gallopin 1996; Phahalad and Bettis 1986; Van Bellen, 2005). This approach promotes the enhancement of public governance (Ministry of Planning 2018) by developing guiding actions (Odum 1996; Prahalad 2009) that go beyond managing only inputs and/or products by the government.

The data sources were processed through differentiated and specific statistical analysis, considering their relevance and significance. This statistical method combines the results of successive indicators moving on to topics, dimension until reaching the final city maturity level.

Mapping the databases to obtain quality information is a critical process that involves several challenges, including:

- Accessibility and data manipulation problems, requiring advanced skills for extraction;
- Data Lake coverage for all Brazilian cities;
- Granularity, often, the database information available is not aggregated at the municipal level;
- Accuracy, which involves data correctness. Some databases contain unreliable information, with clearly incorrect or even incomplete data;
- Temporality and updated data. Some databases lack regular data collection, making it difficult to maintain a historical series. Additionally, due to the lack of open data sources in Brazilian cities, primary data collection was conducted through self-declared questionnaires.

The adaptation of the Smart Sustainable Cities Maturity Model (ITU Y.4904) and the Assessment framework for digital transformation of sectors in smart cities (ITU Y.4906) to the inteli.gente platform has contributed to the maturity level identification of each city, and it should be used in their planning and the public policies development. It has also enabled the construction of a customized system for management and governance, with strategies, indicators, and evolution goals. This allows actions that are aligned with the local context and promotes the exchange of experiences among city administrations. Always prioritizing people, technology should be used as the way to achieve these goals.

The Table 10.1, presents the differences between the SSCMM-ITU model and the model used by the inteli.gente platform. A fourth dimension, Institutional capabilities, was created to accommodate the 5 vertical axes of the SSCMM-ITU model. This dimension provides information and knowledge on public policy actions, facilitating decision-making and priority management in execution plans. The Brazilian weakness in this dimension also contributed to the necessity for the creation of two new maturity levels assigned to the inteli.gente platform.

Another highlight of the inteli.gente platform was its digital format, open accessibility, and free availability, along with customized diagnostics for all 5570 Brazilian cities.

Out of the 91 indicators used in the SSCMM-ITU, the inteli.gente platform maintains adherence and similarity with approximately 70% of them. Regarding the SDG's (sustainable development goals) has approximately 90% adherence, and there is also 40% adherence and similarity with the indicators from the ISO 37120, 37122, and 37123 series.

It is important to emphasize that the objective of the inteli.gente platform is not to rank cities but to provide diagnostics that offer policy direction for smart cities, taking into account their diverse characteristics (CTI/poli.TIC 2020; Manual de Referência 2023).

Accessibility features were implemented in inteli.gente platform since April/2023. It includes navigation in Libras (Brazilian Sign Language), and the WCAG (Web Content Accessibility Guidelines) standards.

10.3.2 Technical Issues

The inteli.gente platform framework, presented in Fig. 10.1, encompasses 113 indicators, distributed across 4 dimensions. The fourth dimension, institutional capabilities, focuses on municipal public management. Additionally, the inteli.gente platform defines characterization indicators that provide sociodemographic, digital transformation, and institutional governance information. The 113 indicators are distributed as follows: 31 for the economic dimension, 12 for the environmental dimension, 31 for the sociocultural dimension, 14 for the institutional capabilities dimension, and finally, 25 indicators are used for characterization.

In Fig. 10.1 are shown the topics (theme areas) related to each dimension. Among these topics, 5, 10, and 8 are associated with the environment, economic, and sociocultural dimensions, respectively. Some topics are repeated across dimensions, such as solid waste and water and sanitation. However, their analysis may differ depending on the dimension in which they are addressed.

The topics covered by the inteli.gente platform also include public policy themes through the institutional capabilities dimension. These topics can be used to guide and understand the indicators in line with their intended purposes. For such, the characterization of a city is defined through its sociodemographic, digital transformation, and institutional data.

The knowledge base and information for each topic are obtained through metrics, statistical analysis, and adaptations based on Odum's organizational theory (1996) and the principles of Bellagio (1997). The data sources come from official regulatory bodies (secondary sources) and/or municipal managers (primary sources).

Table 10.1 Differences between MMSSC—ITU versus inteli.gente

Items	SSCMM-ITU	inteli.gente platform
Multidimensionality	Divided into 3 dimensions Economic, environment, sociocultural	Divided into 04 dimensions Economic, Environment and Sociocultural and Institutional Capabilities. Three new components were built for characterization (a) Sociodemographic, (b) Digital transformation and (c) Institutional; for the recognition of municipalities in territorial arrangements for the transformation and management of the municipality
Level	05 levels	07 Levels, being Level 1—Adhesion, Level 2 Engagement. The other levels 3 to 7 follow the SSCMM-ITU
Axes	05 axes	Created the Dimension of Institutional Capacities that has as topics the 5 verticals of the SSCMM-ITU
Characteristics/Properties/Attributes	CORE: key indicators that all cities should consider when performing the maturity assessment. It is recommended that target values be achieved for all key indicators listed at a given level for cities to claim that they have reached that level	They are relevant indicators to provide essential information to discriminate the technological evolution, the urban structure, and the ICT for a smart city. They follow the evolution of the city's performance and reflect the changing conditions of the economic, sociocultural, and environmental dimensions of the model. Still, they follow an evolutionary logic to drive the diagnosis of urban infrastructures and advances in ICT

(continued)

Table 10.1 (continued)

Items	SSCMM-ITU	inteli.gente platform
	Additional: indicators that cities could consider when developing their own maturity assessment plan and when executing the maturity assessment	For the reality of the Brazilian expansion of the model, they are indicators with attributes in technology and innovation that guide actions and public policies in the city. They are indicators that complement the information of the Core indicators in each theme and/or topic and still meet the needs of services and applications with the use of ICT
	Weights: can be used to reflect their degree of importance in the digital transformation of sectors. It can be used to define the weight of each indicator(s) should determine the weights of all aspects and key areas	High relevance: these are indicators directly linked to the thematic areas of public policies, applied to the themes of each dimension of sustainable development. They report on the adequacy of the city's urban and ICT infrastructure. Average relevance: these are indicators of intermediate relevance, which allow the diagnosis of the evolution of ICT solutions and the improvement of the urban infrastructure available in the city. Low relevance: they are considered less relevant indicators and, in the diagnosis, favor the sustainable digital transformation and the provision and offer of services, solutions and integrated applications in the city
Source	Secundary bases	Secondary base; primary basis of collection by self-declarable form with the public

Thus, the indicators descriptive analysis in the inteli.gente platform provides opportunities and challenges for improving the services' delivery to the population in terms of digital transformation, transparency, and quality.

From 4 October 2021 to December 2022, only 14 Brazilian cities had joined the inteli.gente platform. These cities represent 12% of the total Brazilian population, with representation from out of the 5 Brazilian geographic regions, as shown in Fig. 10.2. The

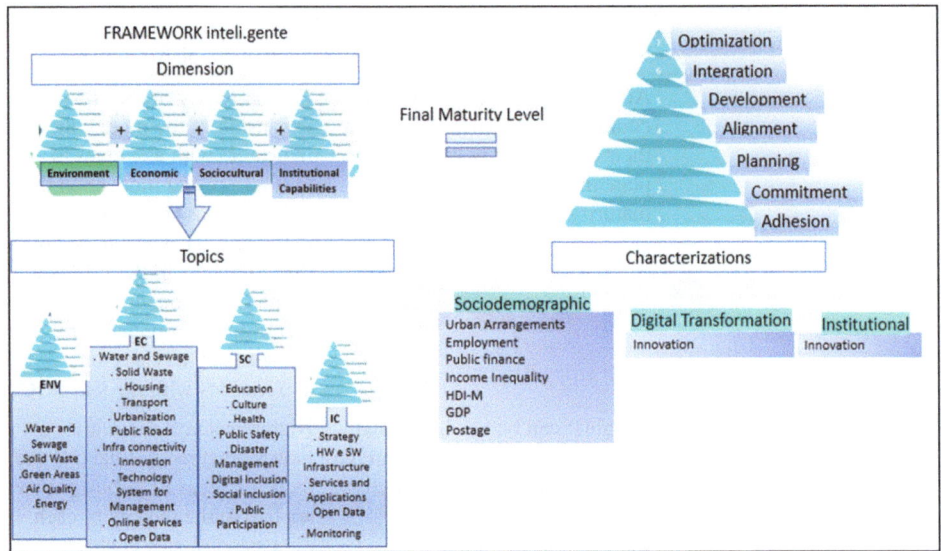

Fig. 10.1 Framework inteli.gente platform (Manual Referencia 2023)

distribution of Brazilian cities is not uniform, with the Northeast, South, and Southeast regions having the highest number of cities, with 1794 cities, 1668 cities, and 1191 cities, respectively. Meanwhile, the Mid-West and North regions have the smallest number of cities, with 467 and 450, respectively.

The low adoption of cities on the inteli.gente was due to difficulties in accessing the online platform and understanding the indicators based on primary data, which relies on responses from municipal managers. This analysis emphasized the need for improvements and changes in the indicators' dependence on primary data, which have been implemented since January 2023.

The average maturity level of Brazilian cities considering sustainability dimensions was at the planning level. This level is defined both in the inteli.gente platform and the SSCMM-ITU. However, in terms of institutional capabilities dimension, the average maturity level achieved was engagement, indicating that municipalities are seeking alternatives for implementing digital and sustainable transformation actions.

The low score in the institutional capabilities dimension highlights the need for formalized planning and strategies for digital transformation, involving municipal managers and policy-makers. The results obtained in this dimension also indicate a need for readjustment of indicators and their metrics, aiming to improve the accuracy of the information being implemented in the inteli.gente platform.

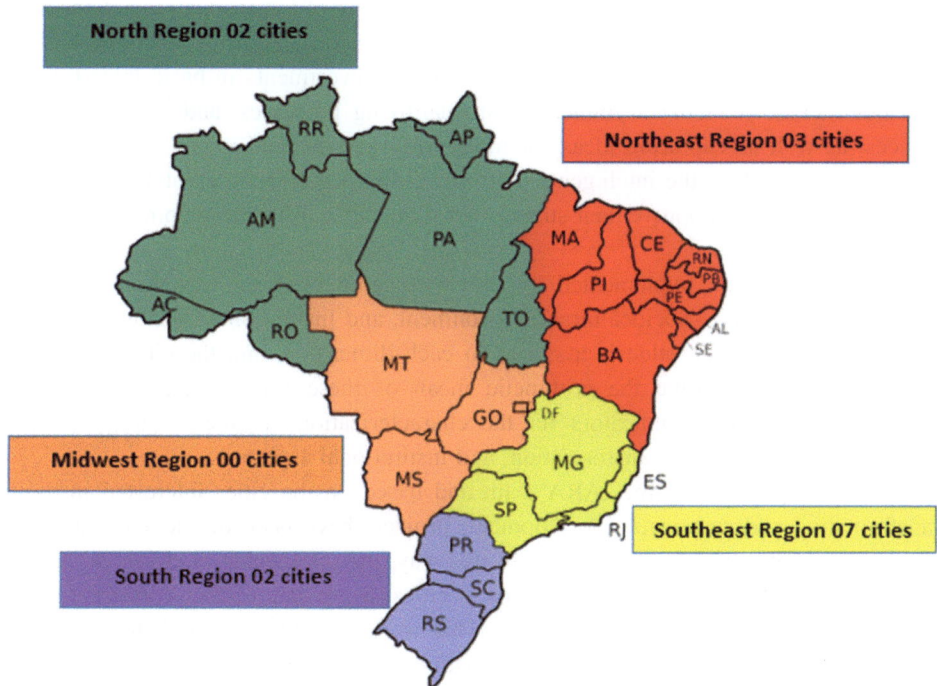

Fig. 10.2 Brazilian geographic regions map and the number of cities adhering to the inteli.gente platform

10.3.3 Discussion

The diversity in Brazil has led to the need for contextualization of recommendations regarding new levels of maturity, with the main objective of encompassing the reality of Brazilian cities and their trajectories, as well as broadening the analyses scope from municipal to national level (Paseto et al. 2021).

As already mentioned, the inteli.gente platform has introduced two new levels. In the first level, called Adesion, the city acknowledges some of its main issues and understands the challenges of its digital transformation. This level is characterized by asymmetries in infrastructure, service provision, and public facilities. The degree of digitalization related to services and processes is low. Cities at this level have the lowest results for sustainable development and ICT indicators, indicating the need for improvement in economic, environmental, and sociocultural aspects. The diagnoses provided by the inteli.gente platform at this level propose the development of public policies that promote the benefits of public management in people's lives and improve sociocultural, economic, and environmental indicators.

In the second level, called Commitment, the beginning of transformation actions in the city can be observed. Sectoral goals emerge for the incorporation of technology. There are actions to identify the most prioritized areas for investments in basic infrastructure. Priorities include reducing asymmetries, standardizing processes, and digitizing public services, as well as capturing ideas for monitoring and evaluating public policies. The diagnoses provided by the inteli.gente platform at this level proposes offering tools and pathways for cities to improve their strategy for sustainable urban development and digital transformation.

Another adaptation implemented by the inteli.gente platform was the use of four dimensions: Economic, Sociocultural, Environment, and Institutional Capabilities. In this model, a maturity level value is assigned to each dimension, and the city's final score is obtained by calculating the arithmetic mean of these four evaluated dimensions. The platform also used indicators for the characterization of cities in terms of their sociodemographic, digital transformation, and institutional aspects.

With the experience of the IARA team and based on the cities interested in digital transformation and sustainable development, projects have been developed that handle large volumes of city data customized. In this way, the inteli.gente platform expands its range of diagnostics and recommendations to cities, using data science (prescriptive and predictive analyses) in its public policies. This work encompasses everything from the necessary infrastructure for digital transformation and urban development to the training and digital literacy required for solutions and services that improve the quality of life of citizens and city management.

The low level of adoption of the inteli.gente platform by municipalities during the period from November 2020 to November 2021 was due to difficulties in online access and filling out responses by municipal managers. This evidence led to changes in data collection starting from January 2023, with the use of secondary indicators. As a result, the inteli.gente platform has provided an estimated level of maturity for all 5,570 municipalities.

10.4 Conclusions

Based on the experience gained in the development of the inteli.gente platform, the following aspects have been highlighted in the version released in 2023:

- Reduced dependency on information and indicators derived from primary sources;
- Improved metrics for the institutional capacity dimension to adapt to the concept of decision-making through multicriteria analysis (Saaty 1984);
- Identified challenges and opportunities across dimensions, applicable to the management and governance needs of cities for digital transformation;

- Enhancement of the data lake formation with 113 customized indicators for each of the 5570 Brazilian municipalities;
- Resumption of meetings with experts to enhance the tracks of challenges and opportunities and explore possible AI applications.

As future implementations, it is envisioned to provide maturity level diagnostics and customized recommendations for all 5570 municipalities in an estimated manner. This ongoing improvement of the inteli.gente platform demonstrates its effectiveness as a valuable tool for the construction, updating, management, and governance of public policies at all levels of government (federal, state, and municipal).

10.5 Revision Questions with Answers

Q1: What motivated Brazil to adopt additional levels of maturity in its smart city maturity model?

A1: The two levels of maturity called Adhesion and Commitment, were added to the Brazilian model to include less advanced cities in terms of sustainable development and digital transformation at inteli.gente platform.

Q2: Regarding the dimensions adopted in the mentioned evaluation models, what is the main difference between the Brazilian model (inteli.gente platform) and the SSCMM—ITU model?

A2: In the Brazilian model, the dimension of Institutional Capacities was added as a tool for evaluating public policies applied to the development of smart and sustainable cities.

Q3: What is the importance of the inteli.gente platform for policymakers?

A3: The inteli.gente platform is a digital, open, and scalable tool that enables not only diagnostics and recommendations but also prescriptive and predictive analyses of maturity levels, fostering the exchange of experiences with other countries in similar contexts.

Q4: What was the latest modification to the inteli.gente platform?

A4: The main modification in 2023 was to estimate a maturity level for each Brazilian city. With this, public agents can make changes in the platform based on this information and improve their maturity level.

Acknowledgements Thanks to MCTI for the possibility of research, to CTI for having started this work (2020–2021), to IARA for continuity (from November/2021 on board), to RNP (2020 on board) for the development of the digital platform and to ANATEL for allowing us to disseminate and present knowledge and new knowledge with SG 20 of SSCMM—ITU.

References

Bibri SE, Krogstie J (2017) Smart sustainable cities of the future: an extensive interdisciplinary literature review. Sustain Cities Soc 31:183–212

Carta Brasileira para Cidades Inteligentes—CBCI (2020). Accessed from Projeto ANDUS—Ministério da Integração e do Desenvolvimento Regional (www.gov.br)

CTI/poli.TIC (2020) Centro de Tecnologia da Informação Renato Archer/Laboratório de Instrumentos de Políticas para Tecnologia da Informação e Comunicação. Modelo de Maturidade de Cidades Inteligentes Sustentáveis Brasileiras. Campinas, SP

Duncan JW et al (1998) Competitive advantage and internal organizational assessment. Acad Manag Exec 12(3):1998

Gallopín G (1996) Environmental and sustainability indicators and the concept of situational indicators. A system approach. Environ Model Assess 1:101–117. https://doi.org/10.1007/BF01874899

International Institute for Sustainable Development (IISD) (1997) Complete Bellagio principles. IISD, Winnipeg, CA. Accessed from https://www.iisd.org/measure/principles/progress/bellagiofull.asp

International Telecommunications Union—ITU-T Y.4904 (2019) Smart sustainable cities maturity model. ITU, Geneva, SWZ. Accessed from https://www.itu.int/rec/T-REC-Y.4904-201912-I

International Telecommunications Union—ITU-T Y.4906 (2019) Assessment framework for digital transformation of sectors in smart cities. ITU, Geneva, SWZ. Accessed from https://www.itu.int/rec/T-REC-Y.4906-201907-P

Lejano Rl (2012) Parâmetros para a análise de políticas: a fusão de texto e contexto. Campinas, SP: Arte Escrita

Manual de referência (2023) Coleta e metrificação de dados para os indicadores da Plataforma inteli.gente: transformação digital para as cidades brasileiras versão 03/Ministério da Ciência, Tecnologia e Inovação; Inteligência Artificial Recriando Ambientes—IARA—ICMC/USP. 1. ed. São Carlos: ICMC/USP, 2023. 171 p.; ISBN 978-85-87837-41-7

Ministério do Planejamento, Desenvolvimento e Gestão (2018) Catálogo De Indicadores PPA 2016—2019. Accessed from http://www.planejamento.gov.br/assuntos/planeja/plano-plurianual/publicacoes/catalogo-de-indicadores.pdf

Nidumolu R, Prahalad CK, Rangaswami MR (2009) Why sustainability is now the key driver of Innovation. Harvard Bus Rev. Retrieved from https://hbr.org/2009/09/why-sustainability-is-now-the-key-driver-of-innovation

Odum HT (1996) Environmental accounting: EMERGY and environmental decision making. Wiley, New York

Paseto L (2021) XI Seminário em TI do PCI/CTI—2021 Infraestrutura: universalidade nas cidades brasileiras

Prahalad CK, Bettis RA (1986) The dominant logic: a new linkage between diversity and performance. Strateg Manag J 7:485–501. https://doi.org/10.1002/smj.4250070602

Saaty TL (1984) Decider face à la complexité, "Une approche analytique multicritère d'aide à la dècision", tradução de Lionel Dahan. Paris. ISBN2-7101-0491-1, pág. 18 à 120

Van Bellen HM (2005) Indicadores de sustentabilidade: uma análise comparativa. FGV editora

Yigitcanlar T et al (2019) Can cities become smart without being sustainable? A systematic review of the literature. Sustain Cities Soc 45:348–365

Luísa Paseto Researcher Coordinator of the inteli.gente platform for MCTI and SG20 Q7 SSC IoT and Smart Cities, Coordinating Researcher IARA Science—Smart Cities. Post-doctorate in Management Systems for Digital Transformation and Sustainable Development. PhD in Management Systems and Sustainable Development, Economist and Business Administrator, with specialization in Planning and Marketing and in Business Ethics. Ambassador of the Brazilian Charter of Smart and Sustainable Cities, collaborator in the Brazilian Chamber Cities 4.0, GIE of Environment and Climate Change—AHK / Brazil, and in the Portuguese Society of Rural Studies.

Márcia Regina Martins Martinez Graduated in Mathematics and Systems Analysis and specialist in Industrial Quality and Smart Cities. Specialist Researcher for inteli.gente platform—IARA. Member of the Technical Group for the Study of Indicators of the Brazilian Chamber of Cities 4.0. Acting in technological development and R&D&I projects, public policies, and development of methodologies. CTI Renato Archer/MCTI Scholarship Researcher, Evaluation of Annual Statements of the Informatics Law (AvalRDA), Certification of National Software Technology (CERTICS); Brazilian Public Software (SPB); National Support Program for Administrative and Fiscal Management of Brazilian Municipalities (PNAFM).

Ricarda Carolina Rende A specialist in regulation and works in the technical advisory department of the Brazil Telecommunications Agency (Anatel). She holds a bachelor's degree and a master's degree in electrical engineering from the Federal University of Uberlândia (UFU) and the National Institute of Telecommunications (Inatel), respectively. In addition, she has completed specializations in public administration law from Unisul, as well as in telecommunications networks and IoT from Inatel. Since 2020, she has been the alternate head of the delegation representing Brazil in the ITU-T. She has shown a keen interest in the study of IoT and smart cities, with a special emphasis on platforms and data monetization.

Rodrigo Barbosa Paula Bachelor's degree in Communications Engineering—Instituto de Ensino Superior FUCAPI—*CESF* and specialization in Digital TV from Federal University of Amazonas—UFAM. For 18 years has been working as a regulation technician at the National Telecommunications Agency—ANATEL and since 2020 has been acting as a delegate in Brazil's representation to ITU.

Andre Carlos Ponce de Leon Ferreira de Carvalho Is professor and the current dean of the Institute of Mathematics and Computer Science in the University of São Paulo, Brazil. His main research areas are data science, artificial intelligence and machine learning. He was the vice president of the Brazilian Computer Society from august 2019 to July 2023. He is the director of the IARA Applied Artificial Intelligence Research Center and is the current the chair of the Machine Learning and Data Mining, in the Artificial Intelligence Technical Committee of the International Federation for Information Processing (IFIP), established in 1960 under the auspices of UNESCO.

Chinese Smart City Standards

11

Ziqin Sang and Wenying Du

Abstract

An effective and interoperable smart city platform (SCP) can break the vertical information silos and build a digital foundation where communication and information sharing between different IoT platforms are possible. Recommendations ITU-T Y.4200 and Y.4201 are international standards mainly developed by Chinese experts from ITU-T SG20 to support city stakeholders in developing an open and interoperable SCP. Y.4201 provides a set of high-level requirements defining the functionalities and interfaces of an SCP while Y.4200 specifies the interoperability requirements of each function and interface in Y.4201. This chapter introduced the development process and main content of them.

Keywords

Information and communication technology (ICT) • Internet of things (IoT) • Interoperability • Smart sustainable city • Smart city platform (SCP)

Z. Sang (✉)
China Information Communication Technologies Group, Wuhan, China
e-mail: zqsang@cict.com

W. Du
China University of Geosciences (Wuhan), Wuhan, China
e-mail: duwenying@cug.edu.cn

11.1 Introduction

This chapter aims to introduce the two important and high-level smart city standards mainly developed by China, including International Telecommunication Union's Telecommunication Standardization Sector (ITU-T) Y.4200 "Requirements for the interoperability of smart city platforms" and Y.4201 "High-level requirements and reference framework of smart city platforms". Here the development history of the two recommendations was provided.

The two Recommendations ITU-T Y.4200 and Y.4201 have gone through a long and difficult discussion before being published. The development of them can be divided into three stages, including the initiation stage, the development and review stage, and the approval stage.

(1) Stage I: Initiation stage (October 2015–January 2016)
During the Study Group 20 (SG20) plenary in Geneva on October 19–23, 2015, China Information Communication Technologies Group (Fiberhome) proposed to initialize a new work item on framework and high-level requirements of smart cities and communities in Q6/WP2, and the SG20 Plenary agreed to create this new Recommendation Y.frame-scc.

During the SG20 plenary in Singapore on January 18–26, 2016, Spain proposed to initialize a new work item on Smart Cities Infrastructure Comprehensive Systems for Smart City Management, and the SG20 plenary agreed to create this new Recommendation Y.SC-platform.

(2) Stage II: Development and review stage (January 2016–July 2017)
During the SG20 plenary in Geneva on July 25–August 5, 2016, Y.frame-scc has been improved with Fiberhome's contributions. Meanwhile, Y.SC-platform has been improved and finalized with Spanish contributions.

At the SG20 plenary meeting, the draft Recommendation Y.SC-platform has been determined with the procedure changed from alternative approval process (AAP) to traditional approval process (TAP) as per request of several member states (ITU-T Res.1 2022; ITU-T A.8 2024). It has been assigned a number Y.4454.

At the SG20 opening Plenary in Dubai on March 13, 2017, according to the TAP consultation result of Y.4454 from ITU member states, this Recommendation ultimately failed to pass the approval. After discussion at the meeting, it was agreed that a new Recommedation Y.SSCP "Requirements for interoperability of smart and sustainable city platforms based on a layered model" was created in Q1/20, based on the content of Y.4454, with Jesus Canada Fernandez and Ziqin Sang as editors.

At the subsequent meetings, it has been agreed to divide Y.SSCP into two parts. Considering the main part "High-level requirements and reference framework of the smart city platform" was similar to the ongoing draft Recommendation Y.frame-scc, it was merged with Y.frame-scc, while the remaining part "Interoperability requirements of the smart city platform" was retained in the draft Recommedation Y.SSCP (renamed Y.SCP).

(3) Stage III: Approval stage (July 2017–December 2017)
During the SG20 plenary in Geneva from 4 September 2017 to 15 September 2017, the two draft Recommendations Y.frame-scc and Y.SCP have been finalized with series of contributions from Fiberhome, Spain and Nokia. They have been consented with the assigned number Y.4201 and Y.4200. The two Recommendations passed the AAP Last Call in October 2017, and passed the AAP Additional Review in December 2017 (ITU-T A.8 2024). They have been published in February 2018.

11.2 Background

City platforms provide various services in specific areas. These services are known in some places as verticals and are provided by supervisory control and data acquisition (SCADA) or more complex platforms. Most of the time these platforms are independent and do not share resources, or they do so sparingly. In city platforms, the proprietary systems abound and the possibilities to offer open interfaces (necessary for the integration with city platforms and data sources) are low.

City platforms may offer some integration; however, currently interoperability problems, such as those below, with city platforms and external providers have been identified.

(a) Many city platforms are conceived as data acquisition and processing systems, without additional possibilities such as data analysis or interoperability with other systems.
(b) There is a tendency for mutual dependence among applications, devices and transport networks.
(c) The concept of horizontality, i.e. allowing the data sensors to be shared by the different services, is barely used.
(d) There is not a trend towards standard open interfaces (open platforms). Although there are already open non-proprietary platforms, the concept of horizontality is not considered.
(e) In some cases, application interoperability is limited to a certain level of interface customization.

There is an increasing need for cities and communities to provide services while also reducing costs, resulting in the need to access different data sources.

In this context the introduction of smart city platform (SCP) enables the integration and optimization of vertical platforms and facilitates the exchange of information and resources between vertical platforms. On one hand, the resources and systems used by the vertical platforms supporting the same functions can be pooled and on the other hand, the information that is stored and processed by one vertical platform can be used by the others, enabling the generation of cheaper, more valuable and more complex services.

Vertical platforms can be integrated with the SCP in two ways. They can be deployed inside the SCP or can be integrated using open interfaces, in case the processes and resources required by the vertical platform could not be deployed inside the SCP. The SCP could interoperate with external providers' city platforms, and its interfaces are required to be adapted. The adaptation of these interfaces will depend on the type of platform.

Figure 11.1 shows the different systems in a city context that can be classified as internal systems (mostly related to traditional community services), like waste, lighting, parking, traffic control, etc. and external systems, such as transport companies, ports, airports, buildings, hotels, social networks, etc.

All these systems could give useful information about the state of a city and can be managed through different types of control tools (IoT platforms, SCADA, non-IoT platforms, big data processors, etc.). However, most of the time these control tools, where they exist, are independent, non-standard and without the possibility of sharing resources and data.

The SCP should have access to different information sources, share resources, analyse capacity and coordinate services, usually based on predictive analysis. The concept of horizontality, where information from all kind of sources interacts in order to provide specific services (instead of using its own sensors), is fundamental to the SCP concept.

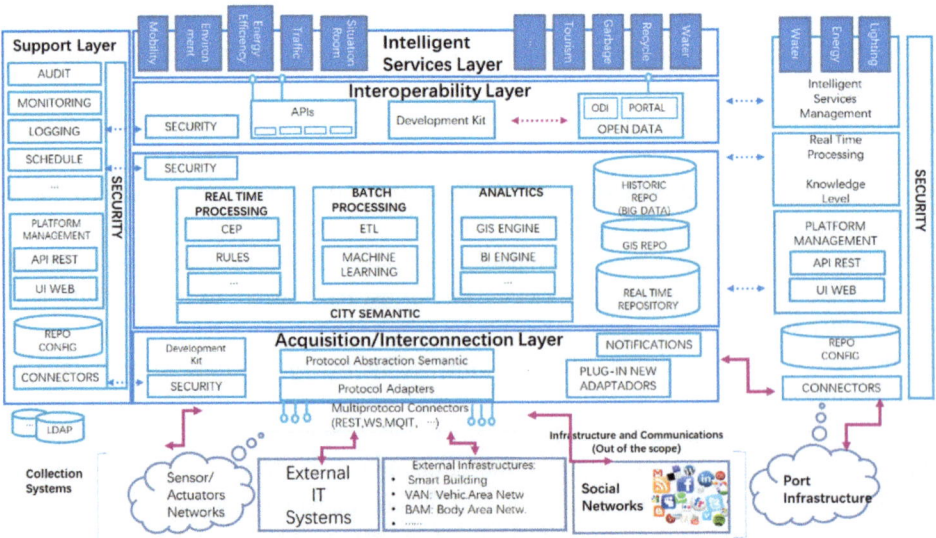

Fig. 11.1 The layout of a smart city platform (UNE 178104 2017)

The range of application cases of the SCP is very diverse, ranging from traffic control during an emergency to the anticipation of demand on a museum due to the arrival of an unexpected number of tourists.

Although SCP can facilitate data sharing and interoperability, there are many parameters influencing the development of smart city standards, including policies of different nations, the sharing and interoperability willingness of smart city enterprises, and the development of smart city correlated techniques, etc.

11.3 Research Methodology: Case Study

The research methodology was developed from the three aspects of strategic, process, and technical level. The case study is the Sharing Application Platform for Government Information System developed by the provincial government of Hubei, China, in accordance to the framework and specifications proposed in Recommendations ITU-T Y.4201 and ITU-T Y.4200.

11.3.1 Strategic Level

The four principles to develop a clear and effective overall smart city strategy are listed as follows.

(1) Overall planning and coordinated development

Starting from the practical issues of the economic and social development of the city, top-level design and unified planning are adopted, including highlighting the key points, point to area, step by step implementation, and gradually deepen, continuously enhancing the effectiveness of implementation and application, and coordinating and promoting the coordinated development of various industries and fields.

(2) Demand oriented, service first

Focusing the construction of smart cities on meeting the needs of the public and enterprises, as well as important issues related to urban development, the continuous improvement of comprehensive service capabilities of the city through informatization are emphasized. "Being able to use, affordable, and well used" is regarded as the fundamental requirement for deepening informatization application and smart city construction, and create conditions for optimizing people's livelihood quality and business environment.

(3) Integration and sharing, innovation and optimization

Fully integrate existing information resources, break down fragmentation, promote intensive construction of smart city infrastructure and application systems, innovate information application systems and government service methods, and continuously optimize project management mechanisms.

(4) Government guidance and diverse participation

Highlight the guiding, encouraging, and supporting effects of policies and funds on the construction of smart cities, strengthen the leading role of market mechanisms and the dominant position of enterprises, mobilize the enthusiasm and initiative of all sectors of society, encourage social forces to participate in the construction, operation, financing, and management of smart cities, and achieve the sustainable development of smart cities.

11.3.2 Process Level

To better procure and manage smart city projects, firstly, special attention should be paid to changing attitudes and improving data governance capabilities. This work means investing more to build a big data center. This is the infrastructure for public data governance, which is very important, but it is necessary to avoid duplicate construction.

The second is to increase the intensity of public data through reforming and improving the sharing of public data resources. In the process of integration, it is necessary to bridge the boundaries between the government, public institutions, public enterprises, and some internet institutions; to create a super app for cities, making it the main portal for mobile services, a hub for application scenarios, a big platform for the government, market, and society, and a convenient and beneficial channel for the people. Currently, multiple cities such as Beijing, Shanghai, Guangzhou, and Shenzhen are building super apps, greatly benefiting the public and enhancing data governance capabilities.

The third is to establish strict systems to ensure the openness of public data resources. Among them, it is necessary to handle the relationship between public security and public data openness, the horizontal and vertical relationship in the development and utilization of public data resources, and the relationship between public data openness and protecting personal privacy. Currently, relevant documents on personal information protection have been issued in China, and finding a balance between public data security and personal data security is a new topic.

Fourthly, in practice, explore opening up some public data resources through market transactions to achieve a win–win situation. Currently, public data trading has received policy support from some local governments. The transaction of public data resources has increased fiscal revenue and promoted enterprises to further utilize public data for the benefit of society. At the same time, it is necessary to promote the free access of some public data to society. There is significant exploration space and new topics in these areas.

11.3.3 Technical

11.3.3.1 Overview of SCPs

An SCP is designed to provide means for enabling new smart city applications to be rapidly created, deployed, extended and managed. An SCP directly integrates city platforms and systems (i.e., SCP functions), or through open interfaces between the SCP and external providers, to offer urban operation and services supporting the functioning of city services, as well as efficiency, performance, security and scalability. As illustrated in Fig. 11.2, an SCP has the following SCP functions:

- services support functions,
- interfacing functions,
- data/knowledge functions,
- acquisition/interconnection functions, and
- security and management functions.

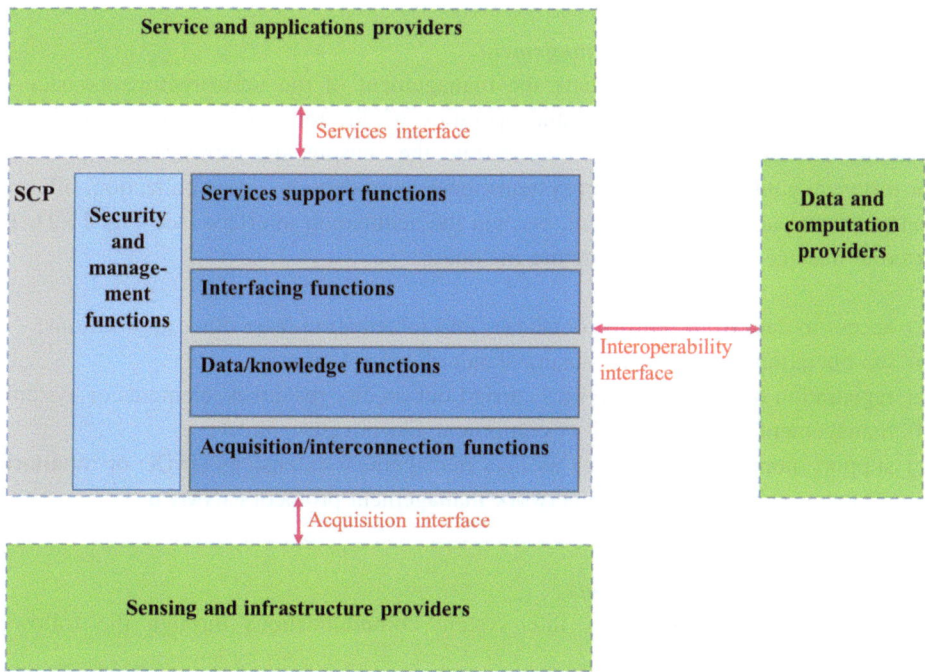

Fig. 11.2 Overview of an SCP and external systems/platforms (ITU-T Y.4201 2018)

The external providers include:

- services and applications providers,
- data and computation providers, and
- sensing and infrastructure providers.

11.3.3.2 High-Level Requirements of SCPs

(1) Comprehensive and Updated Repository of City Information
The SCP is required to handle a comprehensive and updated repository of city information through the following functions:

(a) hosting a freely accessible catalogue of standardized data on the city;
(b) enabling cross-city overviews using this data;
(c) facilitating the integration of data with varying levels of latency;
(d) providing open and standardized interfaces for external providers to access this data;
(e) enabling security and integrity of data, as well as security of users.

(2) Infrastructure Life-Cycle Management
The SCP is required to coordinate the management of the infrastructure, considering secure and multi-user monitoring and operation of different city resources, elements or systems. The SCP is required to orchestrate the sensing and infrastructure providers, for instance, public transportation systems, energy production, waste collection, off-street and on-street parking management, etc. via the acquisition interface (see Fig. 11.2). The external management of infrastructure includes:

(a) access to data from sensors, databases and information from other applications;
(b) the operation of city sensors using standardized solutions;
(c) registration of different activities carried out on city resources, elements or systems;
(d) management of maintenance of equipment and infrastructure;
(e) support monitoring tools such as Java management extensions (JMX) or monitoring protocols such as the simple network management protocol (SNMP).

(3) Inter-System Communications
The SCP is required to allow inter-system communications through the following functions:

(a) use open and secure application programming interfaces (APIs) and standard protocols to communicate between applications and with other management systems;
(b) conditional and secure access to emergency services.

(4) Security Support

1) General Aspects
The SCP is required to include appropriate measures to maximize privacy and security of citizens by implementing tools in order to:

(a) Back up critical information.
(b) Guarantee non-repudiation in network security.
(c) Support anonymized data.
(d) Guarantee security and integrity of data.
(e) Support authentication and authorization.
(f) Control access to the platform and to all the elements accessed through it.
(g) Support confidential communications.
(h) Ensure confidential access to data, ensuring that each agent only has access rights to the data assigned to it.
(i) Define and manage security policies.
(j) Maintain repositories of existing users registered by local authorities.

2) Profiling
The SCP is recommended to ensure the privacy or security of the data stored or managed by the city, particularly in environments with shared resources. Furthermore, different access profiles can access different types of groups of data, thereby avoiding their misuse.

The SCP is recommended to guarantee the secure exchange of data along the platform from the physical devices to applications.

The SCP is recommended to allow for the definition of different profiles (such as administrator, external providers, citizens) with appropriate access rights, authorizing or denying access to the different applications and defining the required privileges for operating with identified sets of data.

Management of roles/permissions is recommended to be established at least for three security levels:

(a) Data access: limiting the information that can be viewed by each user. For example, a user of a specific service only has access to information corresponding to its service.
(b) Access to the elements of the platform: limit access to reports and dashboards configured in the platform. For example, a user of a service is only able to access reports defined by the data corresponding to their role.
(c) Functions: limited actions available to a certain user according to their profile. For example, a user of a report may determine which reports or objects final users have access to.

(5) Maintenance Support

The SCP is required to include appropriate measures for its adequate maintenance.

The SCP is recommended to support preventive maintenance and corrective maintenance.

(6) Showcase Support

The SCP is required to include the functionality of showcasing the status of its infrastructure and services. The presence of information includes:

(a) Status of city infrastructure:
- engineering infrastructures, such as energy supply, water supply and drainage, transportation, logistics, communication, environment and disaster prevention;
- social infrastructure, such as administration, welfare, health, education, science, culture, sports, and entertainment.

(b) Information of services (ITU-T Y.4903 2022):
- economy aspects, such as employment, trade, productivity and innovation;
- environmental aspects, such as air quality, water and sanitation, noise, soil and biodiversity;
- societal and cultural aspects, such as education, health, safety, housing and culture.

(7) Controls of Processes

It is recommended that the SCP supports controlling the following internal and/or external processes:

- consumption analysis, warnings and trends, etc.;
- cost allocation;
- sustainability (efficient use of facilities, greenhouse gas emissions, etc.) and reliability of infrastructures;
- optimization of processes and planning;
- crisis dashboard.

The SCP is also recommended to support the generation of operating reports.

(8) Support for Decision Making

The SCP is recommended to support decision-making processes to improve city resilience by providing:

- simulations based on current and past information;
- assessment and deployment of action plans in complex scenarios;

- predictive modelled analysis of the city;
- data mining and statistical analysis;
- integration with business intelligence systems and tools.

(9) Real Time Dissemination of Public Information

The SCP is required to enable transmission of open, reliable and quality information, on a constant basis and free from interruptions, in standardized formats to be accessible from multiple ICT devices.

This information is applicable to the following scenarios:

- end-user citizen services (the information society);
- external services and applications;
- other administrative public services;
- services for purposes of accountability (transparency).

(10) Resiliency

The SCP is recommended to:

- Ensure the ongoing operation of services according to established service level agreements (SLAs). These services may require availability 24X7 and a service level of over 99.9% annually. Providers may offer solutions complying with these requirements.
- Guarantee disaster recovery with limited recovery time objectives (RTOs) and recovery point objectives (RPOs).

(11) Interoperability

In order to ensure an adequate interoperability service, the following aspects are required:

(a) **Interoperability with different technologies**: capability of supporting different technologies for capturing information and communications standards, as well as internal/corporate and/or external information systems.
(b) **Performance**: ability to handle a large number of devices, services and processes efficiently.
(c) **Scalability**: ability to increase processing, interconnection and storage capacities without needing to change the architecture.
(d) **Robustness and resilience**: capability to continue while facing problems.
(e) **Security**: guarantee of security and reliability.
(f) **Extensibility**: adaptability to meet new needs.

11.3.3.3 Main Functions of SCP

Main functions of SCP can be described in detail (shown in Fig. 11.3). Each function offers capabilities; these are detailed below:

- Acquisition/interconnection functions: offer mechanisms for capturing data from the collection systems. They also enable interconnection with other external providers that only use data. These functions provide other capabilities such as:
  - Abstract: This allows for the gathering and refining of data from devices for desired processing in an agnostic manner.
  - Tagging: This allows for the received information to be transformed to the data structures defined in the platform. It allows for the source of the data received to be identified by adding tags including identifier and registration information.
- Data/knowledge functions: support the processing of data. They receive data from both the acquisition/interconnection functions and the interfacing functions and include functionalities which enable the movement of large amounts of data, and data processing and analysis functionalities to generate new datasets or modify/complete existing ones. Other functionalities offered are:

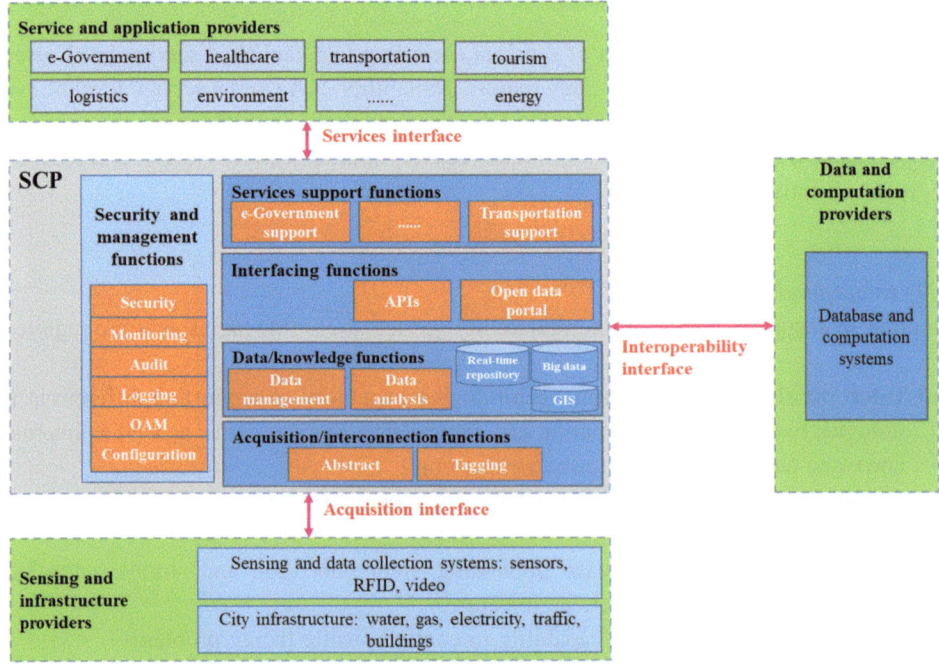

Fig. 11.3 Reference framework of an SCP (ITU-T Y.4201 2018)

- Data management: This provides a systematic way to create, retrieve, update and manage data.
- Data analysis: This provides the capacity to analyse the data.
- Big data: This provides the capacity to store and work with large amounts of data.
- Geographic information system (GIS): This provides the capacity to work with geospatial information.
- Real-time repository: This provides the capacity to work with information in real time.
• Interfacing functions: provision of services for smart cities. They also offer open and standardized interfaces for the data/knowledge functions and the services support functions, which are compliant with the security policies. Other functionalities that are provided are:
  - APIs that allow access to the service provided by the SCP.
  - Open data portal that allows data to be made publicly available.
• Services support functions: These support various services connected through interfacing functions and the APIs supplied. Applications can be running on the platform or can be other external services that publish or use the information. Some services support functions that may be offered are:
  - e-Government support: These services are enabled to match electronic public administration and other government needs.
  - Transportation support: These services manage transport information and allow the traffic of the city to be managed.
• Security and management functions: These functions provide horizontal support to the other functions by offering services such as audits, monitoring and security. Some of the functions that may be offered are:
  - Security: This provides security mechanisms such as authentication, authorization, ciphering, etc.
  - Monitoring: This collects platform operation information.
  - Audit: This registers who has accessed sensitive information.
  - Logging: This traces the execution of the applications and/or the systems of the platform.
  - Configuration: This allows access to the configuration of the systems and to change the execution parameters.
  - Operation, administration and maintenance (OAM): Processes and tools that allow the platform to be operated, managed and maintained.

11.3.3.4 Reference Points for SCP Interoperability with External Providers

Figure 11.3 shows interfaces which enable the communication between the functions (red arrows). Details of these interfaces are described in the following list:

- **Acquisition interface**: This interface with the SCP enables information collection from the external elements.
- **Interoperability interface**: This interface with the SCP enables communication with external data providers and the third-party computation systems.
- **Service interface**: This interface with the SCP enables application to application access to support functions provided by the SCP.

(1) Acquisition Interface

It is required that the acquisition interface of the SCP complies with the following technical characteristics:

- network access and sensor technology independence, i.e., compatibility with different network access and IoT/machine to machine (M2M) protocols;
- support for open protocols and protocol translation, to ensure that the platform remains independent from the complexity of devices and enables access to sensors/actuators from different manufacturers;
- access to sensors/actuators: sensor/actuators information is collected through a transport network;
- support of security and monitoring functions;
- discovery and access to IoT/M2M applications;
- support of identification and naming of devices and applications.

(2) Interoperability Interface

It is required that this interface enables the interoperability between the SCP and external systems and allows access to data, information and services that are stored or provided by the SCP.

It is required that this interface includes authentication and authorization aspects that allow control access to the functions. The authorizations will be granted according to the terms of use.

It is recommended that this interface enables internal access to the data management and basic capabilities offered by the acquisition/interconnection functions:

- to enable access to the metadata of sensors registered in the platform;
- to implement authorization and authentication functions for the different available actions;
- allowing real-time collection of data generated by a sensor or group of sensors.

It is recommended that this interface:

- allows operation with data sets. The interface should provide functionalities that allow mathematical operations on the data.
- The interface allows extraction and analysis processes. The interface should provide functionalities that allow analysis based on big data, that is, analytical capabilities to transform data into valuable information.
- It is required that this interface enables internal access to the information management offered by the service support functions through application programming interfaces (APIs) providing different data access modes, including push (subscription and notification) and pull (request and response).

(3) Service Interface

It is required that the service interface offers APIs and other tools such as a development kit and open data portals, which will be used to implement the services delivered to clients.

It is required that this interface:

- provides a secure access to the APIs, development kit, web portal, etc.;
- is based on open APIs (including the provision of an API manager) that can be used by the internal or external applications;
- follows the overall API representational state transfer (REST) trend.
  It is recommended that this interface:
- makes available a web portal that can be used from the services offered;
- supports different data access modes, including push (subscription and notification) and pull (request and response);
- provides the mechanism necessary to adapt the communications to the data models and semantics.

11.3.4 Discussion—The Fringes

The Recommendations ITU-T Y.4201 and ITU-T Y.4200 have already made demonstrable impacts and devised innovative strategies to enhance sustainability and resiliency at the city level. The following is the use case that demonstrate the impacts, innovation, and viability that the Recommendations ITU-T Y.4201 and ITU-T Y.4200 have had in China. The use case is the Sharing Application Platform for Government Information System in Hubei, China.

Over the years, China has emerged as one of the world leading countries in smart city development. The country has over 500 smart cities, each is looking to leverage data-driven technologies to develop innovative solutions and applications that would enhance their overall energy efficiency, improve the quality of social services as well as the overall livability of the cities.

In 2015, the provincial government of Hubei developed a Sharing Application Platform for Government Information System. This information sharing application platform is developed in accordance to the framework and specifications proposed in Recommendations ITU-T Y.4201 and ITU-T Y.4200. The purpose of this platform is to coordinate captured data and apply them to improve city services.

(1) Impacts and Innovation
Through the Sharing Application platforms of government information systems, many of the isolated data islands have been eliminated. The work efficiency of the city's digital applications has been greatly improved. The sharing of government information through this platform has also encouraged the citizens to leverage those data to develop innovative applications and sustainability solutions such as mobile applications that use open data to predict the patterns of extreme weather.

Since its construction in 2015, the platform has been instrumental in realizing the "digital government" vision of the Hubei provincial government. The platform is now actively involved in improving medical health, maximizing the efficiency of public transportation, monitoring upcoming extreme weather events, improving disaster management, reinventing the education system along with other public services. At present, there are over 700 digital government applications from over 70 government departments running on this platform.

(2) Viability
The impact of the Sharing Application Platform goes beyond the provincial government. In Hubei Province, 17 prefecture-level municipal governments and 63 county-level governments have already started building Sharing Application Platforms using a similar structure of the provincial smart city platform. The provincial-level Sharing Application Platforms are also expected to be connected to other municipal and county-level platforms. Eventually all the platforms will be connected to the central government platform, forming a four-level digital government network, namely, the central government platform, provincial platforms, prefectural municipal platforms and county platforms, that would be the foundation for realizing the country's digital agenda.

This high level of interconnectivity also lays the groundwork for effective digital governance. The government's capability to deliver IoT-based sustainability and resilient strategies will be greatly increased. Thanks to the nominated Recommendations, the specifications, functionalities, and terminologies of each digital platforms will be standardized, allowing data management to be extremely efficient, greatly reducing data friction between different data platforms and promoting the construction of a coherent information society.

11.4 Conclusions

Recommendations ITU-T Y.4201 and ITU-T Y.4200 advance sustainability and resilient in smart cities by providing the blueprint of an open and interoperable smart city platform that is capable of addressing a wide-range of city challenges including but not limited to urban sensing, infrastructure management, climate change, and citizen-centered integrated services. Such platform is the digital foundation for circulating data collected by different sensor networks and other sources and translating them into actionable insights that support city stakeholders in making better decisions.

Most of the existing city platforms are verticals and they are not designed to have the capacity to share data with other external platforms:

- They are developed solely for the purpose of data acquisition and as a data processing system.
- They are not designed for data analysis and do not take interoperability with other systems into consideration.
- The concept of horizontality, i.e. allowing data collected by different sensor networks to be shared among different services and platforms, is barely present if at all.

Therefore, the insights generated from these verticals are usually based solely on the data collected by their respective database. However, cities are facing environmental, economic and developmental challenges that are often of multidimensional in nature. Take transportation as an example, a vertical with sensors networks dedicated to collect and process data for the purpose of relieving traffic congestion may exist. Yet, traffic congestion in city contributes significantly to the city's greenhouse gas emission and has direct impacts on the quality of social services, such as the delivery time of emergency services, as well as the health of citizens.

The two Recommendations offer cities a pathway to overcome this challenge by reinventing the city data platform into an open and horizontal one. Implementing the SCPs brings the immediate benefits of allowing city stakeholders to make better decision. They will have access to insights that are generated from multiple databases instead of a single vertical. For example, with the SCP, city regulators would be able to leverage the data from the 'transportation' vertical to address other sustainability challenges including climate change adaptation, mitigation or monitoring and vice versa.

The implementation of SCPs brings the long-term benefits of reducing the overall number of data centers and increasing energy efficiency for the existing data centers since an effective and open SCPs would eliminate the need to build new verticals for every aspect of a city or other emerging situation. The consolidation of data would also allow new and innovate sustainability solutions and business models to be developed.

In addition, the fact that the two Recommendations are international standards offer another distinctive advantage to prospective city stakeholders who are looking to implement an SCP. With both standards being freely available to anyone, it allows city governments to take a proactive approach to improving sustainability and resiliency in city. Cities are independent from needing to receive technical supports from telecom providers and operators which usually involves a lengthy and costly process, to build an SCP.

11.5 Revision Questions with Answers

(1) What is the high-level requirements of SCPs?

- Comprehensive and updated repository of city information
- Infrastructure life-cycle management
- Inter-system communications
- Security support
- Maintenance support
- Showcase support
- Controls of processes
- Support for decision making
- Real time dissemination of public information
- Resiliency
- Interoperability

(2) What is the main functions that SCP should provide?

- **Acquisition/Interconnection functions:** These provide data capture mechanisms from the collection systems.
- **Data/Knowledge functions:** These support data processing, adding value and transforming information into knowledge.
- **Interfacing functions:** These enable access to information at different levels.
- **Service support functions:** These coordinate all the possible services involved in each action developed out of interoperability functions.
- **Security and management functions:** These provide horizontal functionalities such as audits, monitoring and security.

(3) What are the requirements for interoperability of smart city platforms?

- **Interoperability with different technologies**: capability of supporting different technologies for capturing information and communications standards, as well as internal/corporate and/or external information systems.

- **Performance**: ability to handle a large number of devices, services and processes efficiently.
- **Scalability**: ability to increase processing, interconnection and storage capacities without needing to change the architecture.
- **Robustness and resilience**: capability to continue while facing problems.
- **Security**: guarantee of security and reliability.
- **Extensibility**: adaptability to meet new needs.

(4) Which interfaces could enable the SCP interoperability with external providers?

- **Acquisition interface:** This interface with the SCP enables information collection from the external elements.
- **Interoperability interface:** This interface with the SCP enables communication with external data providers and the third-party computation systems.
- **Service interface:** This interface with the SCP enables application to application access to support functions provided by the SCP.

Acknowledgements Thanks for the invitation from Dr. Leonidas Anthopoulos. We also appreciate the valuable feedback from the participating experts in ITU-T SG20, especially Jesus Canadas Fernandez, Francisco Javier Carvajal Pérez, and Blanca Gonzalez, during the development and review process of Recommendations ITU-T Y.4200 and ITU-T Y.4201.

References

ITU-T Resolution 1 (2022) Rules of procedure of the ITU telecommunication standardization sector. Retrieved September 18, 2023, from https://www.itu.int/dms_pub/itu-t/opb/res/T-RES-T.1-2022-PDF-E.pdf

Recommendation ITU-T A.8 (2024) Alternative approval process for new and revised ITU-T recommendations. Retrieved June 12, 2024, from https://www.itu.int/rec/T-REC-A.8-202401-I/en

Recommendation ITU-T Y.4201 (2018) High-level requirements and reference framework of smart city platforms. Retrieved Sept 23, 2023, from https://www.itu.int/rec/T-REC-Y.4201-201802-I

Recommendation ITU-T Y.4200 (2018) Requirements for the interoperability of smart city platforms. Retrieved Sept 26, 2023, from https://www.itu.int/rec/T-REC-Y.4200-201802-I

Recommendation ITU-T Y.4903/L.1603 (2022) Key performance indicators for smart sustainable cities to assess the achievement of sustainable development goals. Retrieved Sept 28, 2023, from https://www.itu.int/rec/T-REC-Y.4903-202203-I

UNE 178104 (2017) Sistemas Integrales de Gestión de la Ciudad Inteligente. Requisitos de interoperabilidad para una Plataforma de Ciudad Inteligente. Retrieved Sept 21, 2023, from https://tienda.aenor.com/norma-une-178104-2017-n0059471

Ziqin Sang, China Information Communication Technologies Group. Mr. Ziqin SANG joined Wuhan Research Institute of Posts and Telecommunications (WRI) in 1998 after obtaining his Ph.D. in pattern recognition and intelligent system from Huazhong University of Science and Technology, China. He heads a division of innovative research on smart city in the State Key Laboratory of Optical Communication Technologies and Networks since 2010. He has been active in standardization bodies such as ITU-T SG17, SG13, SG5 and SG20, CCSA TC8 and TC10. He served as a member of expert board of a Chinese 863 program of smart city, and a member of advisory board of EU-China Collaboration on IoT and 5G Research (program EXCITING). He was a vice chairman of ITU-T focus group on smart sustainable cities (FG-SSC) and a vice chairman of focus group on smart water management (FG-SWM). He is now a vice chairman of ITU-T SG20 and a vice chairman of CCSA TC10. He led and contributed to ITU-T Recommendations Y.4900, Y.4901, Y.4902, Y.4903, Y.4200, Y.4201 and Y.4216. He is currently a member of editorial board of IET Smart Cities, MDPI Smart Cities and MDPI Standards.

Wenying Du, China University of Geosciences (Wuhan). Ms. Wenying DU acquired her Doctor's Degree in Photogrammetry and Remote Sensing from Wuhan University in 2017, and is currently an associate professor and master's supervisor at China University of Geosciences (Wuhan). She acted as PI in three Chinese national or provincial projects, and participated as a backbone in four Chinese national programs. She published 29 academic papers, authorized 5 invention patents, registered 3 software copyrights, participated in writing 2 academic monographs. She also served as reviewers for multiple academic journals, i.e., Journal of Hydrology, Sustainable Cities and Society. She contributed to ITU-T Recommendation Y.4562 and is now working on another three work items in ITU-T SG20.

Smart City Standardization in France: The Case of Proxi-Produit Project to the DPP Policy

Sandoche Balakrichenan

Abstract

In the context of a smart city, the Internet of Things (IoT) plays a pivotal role in enhancing urban living by integrating various devices and systems to collect, analyse, and act on data in real time. Identification of a particular object in IoT plays a crucial role. With the heterogeneity in IoT identification, Domain Naming System (DNS) infrastructure could play a vital role in ensuring seamless communication and interoperability among the myriad IoT devices deployed across the city. This article explains the process and technicalities involved in designing and implementing a proximity-based citizen service called 'Proxi-Produit'. The Proxi-Produit platform was built on a standard called the Object Naming Service based on DNS, which needed the cross-pollination between two different standards organisations, GS1 and IETF. We conclude by mentioning that the experience in this Proxi-Produit platform could help design and develop the tools and technology for the Digital Product Passport initiative, which will become mandatory in the EU for a range of different products.

Keywords

IoT · DNS · ONS · Smart city · Proximity-based services · Circular economy

S. Balakrichenan (✉)
Association Française Pour Le Nommage Internet en Coopération (Afnic), Guyancourt, France
e-mail: sandoche.balakrichenan@afnic.fr

12.1 Introduction

The European Commission is driving a regulatory framework called the Digital Product Passport (DPP) (REGULATION (EU) 2024) to create transparency around product information and accelerate the transition to a circular economy. The objective of DPP is that products sold in the EU will require an attached unique identifier that includes detailed information about materials used, manufacturing processes, recyclability, CO2 impact, etc. By revealing a product's journey and environmental implications, DPPs will empower consumers to make informed purchasing decisions and pave the way for a greener, more ethical future, thus orchestrating a smart circular economy (Langley et al. 2023) (e.g., in a smart city).

The Internet of Things (IoT) will play a significant role in building a DPP. The IoT paradigm represents connecting anything and everything for communication purposes. IoT enables real-time monitoring of product conditions throughout their lifecycle, improving data security, accuracy, and tracking capabilities.

The product must be associated with a carrier device to resolve the product identifier in an online information catalogue. To communicate on the Internet, a computer requires a carrier device such as a network card; similarly, mobile phones need a carrier device such as a SIM card. Examples of carrier devices in IoT are barcodes, RFID, sensors, NFC, etc.

The carrier devices in the Internet are identified uniquely in the scope of the Internet by identification mechanisms such as IP address, MAC address, domain names, etc. In IoT, we have a plethora of technologies. Each uses its identifiers, naming conventions, namespaces, and discovery process. A problem that arises due to this is that a "thing" using one technology (e.g., Zigbee protocol) will not be able to communicate with a "thing" using another technology (e.g., Wi-Fi). In simple terms, the problem is "interoperability".

The question is: with such a heterogeneity in identifier naming conventions and provisioning structure, will it be possible to communicate between 'things' that use different identifier naming conventions and communication technologies? Two approaches to facilitate interoperability exist: disruptive or evolutionary. There are some disruptive approaches (Amadeo et al. 2014) for IoT. Most tend to be proposals from academia, implemented in a laboratory environment (or) tested in specific use cases.

The evolutionary approach uses existing technologies that have withstood the Internet's operational strains. The Domain Naming Service (DNS) (Mockapetris 1987a, 1987b) is one such technology which has existed since the beginning of the Internet and remains its cornerstone. Even though the Internet evolved with a scale not initially thought of, DNS remains the essential infrastructure for resolving information on the Internet and interoperating between different applications and services.

DNS was conceived to translate human-friendly computer host names on a TCP/IP network into their corresponding machine-friendly IP addresses. Besides translating host names to IP addresses, DNS is used, for instance, by mail transfer agents to find out where to deliver mail for a particular address, as a general mechanism for locating services in a domain using Service (SRV) (Gulbrandsen et al. 2000) records, as a resolution of identifiers that do not have traditional host components through DNS using Naming Authority Pointer (NAPTR) (Mealling and Daniel 2000) resource records etc.

For IoT, overlay mechanisms services exist, such as Object Directory Service (ODS) (ITFIND n.d.), which uses the DNS infrastructure to resolve IoT identifiers (using legacy naming conventions) to their related digital information on the Internet.

This article will share the experience of a predecessor of DPP called "Proxi-Produit" (Proxi-Produit Consortium 2009), which was tested in France. In Proxi-Produit, we tested and implemented an architecture that described how a product identified by GS1 standards (Standards | GS1 n.d.) could be linked to any number of resources on the Internet, wherever they may be, via DNS. The benefits of fusioning two standards for a project like 'Proxi-Produit' are explained in subsection 3.5. The article intends to show that lessons learned from the project could greatly benefit the development of the tools and technology for DPP.

12.2 Background

12.2.1 Context that Led the 'Proxima Mobile' Initiative—A Bouquet of Citizen Based Services

The rise of the mobile Internet is accompanied by the consideration of contexts in which users are required to use the services made available to them. With a significant mobile communications market and a very dense network of innovative Small and Medium-Sized Enterprises (SMEs), France 2010 wanted to develop digital services to support citizens' daily activities. The Proxima Mobile (PROXIMA MOBILE 2013) project was conceived with the dual objective of creating services that are useful to all citizens and stimulating the ecosystem of mobile services.

Proxima Mobile was to enable the creation of a bouquet of services of general interest in fields as diverse as education, law, health, tourism, disability, local life, employment, transport and even sustainable development. Proxi-Produit is one of the Proxima Mobile projects which focused on consumer products. The Proxima Mobile initiative was financed as part of France's recovery plan (MISSION PLAN DE RELANCE DE L'ECONOMIE 2009).

This project, coordinated by the "Délégation aux usages de l'Internet" (DUI) under the supervision of the French Ministry responsible for the digital economy and the Ministry of Research, had a double objective: to offer citizens services that make their daily lives easier and stimulate the development of the mobile services ecosystem. It was also about helping innovative SMEs to develop partnerships with administrations or local communities to develop beneficial services to citizens on a national, even European, scale.

12.2.2 Requirements of Proxima Mobile Service

Specific criteria were retained for the selection of applications and services of the Proxima Mobile services, which include:

Daily use citizen services:
The requirement is to create a platform that enables the possibility to dynamically create new services (commercial, cultural, environmental, etc.) by adding new layers of information. Unlike the usual trend, which pushes marketing information towards the consumer, the requirement is to allow mobile users/consumers to access the product information they need or even essential.

A shared platform for service providers:
The use of the Internet as a source of information by consumers and the rise of "unofficial" product information services are leading brands to organise themselves to obtain the visibility they deserve. Currently, the visibility is provided by individual corporate websites. The requirement is to use a shared channel to diffuse official product information.

Connectivity anywhere/anytime:
The requirement is to allow consumers to be connected anywhere, all the time (in stores, at home, etc.) and to search for contextualised information (geolocation, price comparison, in-store availability, etc.).

Ergonomics—Key for citizen services:
The rise of mobile devices is primarily linked to the simplification of interactions with the user. Touch screens have made the spread of, and access to information easier. These improvements in ergonomics have enabled people who cannot use computers to benefit from mobile services. Thus, the ergonomics of mobile services contribute to reducing the digital divide among disadvantaged households.

Accessible Citizen Services that are profitable and sustainable:
The objective was that Proxima Mobile services must not contain advertising. Rather than a political choice, it was a question of avoiding advertising banners, which occupy an important place on the screen and reduce the "useful" surface area for displaying and consulting content. In addition, mobile advertising remains the least appreciated by users. Accessible and free services enable a broad audience to benefit from it, but also to subsequently develop "premium" versions of said services that provide more features than the free version.

The aim was to use open standards to provide the services so that there is no recurring cost in terms of license, Intellectual proprietary rights, etc.

Innovative methodology:

Beyond mobile terminals, the objective was also to connect everyday objects. With IoT, it will become possible for network users to receive notifications from, and to interact permanently with all the objects present in their immediate environment. In addition to the trade, tourism and transport sectors, energy professions could also be transformed by the contribution of mobile technologies. Indeed, installing sensors on the electrical network and within subscribers' premises (companies and individuals) will enable significant energy savings. In the same way, energy, medical or environmental issues could be the origin of new services connected to mobile terminals.

12.2.3 Standardisation Committee

The basis for Afnic to be invited to be part of the 'Proxi-Produit' project under the Proxima mobile call for proposals is our previous experience developing a GS1 standard called the Object Naming Service (ONS) (Object Name Service (ONS) 2013). A process is followed to create the ONS standard under the GS1 ambit, which is explained in the following paragraph.

An adhoc group was created under the GS1 standards umbrella to develop the ONS standard. The adhoc group completed the work on the ONS requirements. The adhoc group is the initial step in forming a working group for the GS1 standards. A call for action was initiated to ask interested companies to join the Mission Specific Working group which is the next step to develop the GSMP GS1 EPCglobal Federated Object Naming Service. The interested parties could be GS1 organisations, solution providers, manufacturers, distributors and retailers. As ONS uses DNS, Afnic was invited to the group for their DNS expertise.

12.2.4 Proxi-Produit—Project Partners and Activities

As part of the call for Proxima Mobile projects launched by the French State Secretariat for Digital Economy Planning & Development, the "Proxi-Produit" consortium involved several stakeholders in France, including Afnic, the French Internet registry (www.afnic.fr); GS1 France (https://www.gs1.fr/), the organisation which manages the supply chain registry for France and Adenyo, a full-solution mobile marketing and media provider.

Adenyo was involved in Project management, technical development and operation of turnkey mobile solutions. GS1 France positioned itself in the consortium as a guarantor in implementing GS1 standards and in conducting subsequent standardisation work, which may prove necessary for encouraging the publication of information relating to products by brands. GS1 France's objective was to create a standard and open platform allowing citizen-consumers to access content relating to the product. Afnic, as an actor managing the Internet namespace for France and its overseas territories, appeared as an element of trust within the consortium. Afnic also brought its expertise in the development of ONS, which played a role in resolving the product identifier to its resources.

The project planning was in three steps:

- Firstly, designing and implementing the 'Proxi-Produit' platform;
- Secondly, getting multiple stakeholders involved, such as manufacturers, retailers, distributors and consumer associations, and organize them in multiple working groups;
- Thirdly, implementing the outcome of the different working groups and provision the Proxi-Produit database with different product information.

12.3 Research Methodology: The Case of Proxi-Produit

12.3.1 Service and Operational Requirements

The Proxi-Produit project aimed to create proximity services linked to the product through an open and standardised architecture, allowing information owners (industrialists, distributors, and communication organisations) to communicate with consumers/citizens directly on their mobile phones.

On an operational level, the project objective was to democratise the use of mobile services by allowing the creation of a viable ecosystem around access to product information via mobile. On a technical level, the project aimed to develop a distributed architecture of mobile services and applications open to information owners to reach their "target" on mobile. Ethically, the project aimed to prioritise social, practical, and useful services for people with difficult or restricted access to information.

The applications developed as part of the Proxi-produit had the common prerequisites:

- **Openness to all**: allow any individual to access and use these services, as long as they have a mobile phone and an appropriate subscription.
- **Economic viability**: No service, however virtuous it may be, will be able to continue if the value perceived by the user shrinks and the economic model is not sustainable.
- **Interoperability**: this involves guaranteeing the solution's openness (aggregation and distribution) to as many terminals as possible while considering the evolution of mobile platforms and OS.
- **Industrializability**: Able to be developed with all industrial partners and distributors. From its origin, it must be built on an open, standard architecture capable of supporting significant load increases.

12.3.2 Process Level Issues Implementing Proxi-Produit

This section describes the administrative and process issues in procuring and managing information related to Proxi-Produit as the Smart City project.

Too many commercial and proprietary offers:
"Product" information via mobile is already part of the commercial and marketing strategy of manufacturers and distributors. The approach is generally focused on the product and its merits and not on the benefit that the consumer can derive from it.

The aim of these information services is solely to increase consumption volumes and achieve the fastest possible "Return on Investment" (ROI). The industry generally follows two approaches:

- **Product launch**: This mercantile approach is generally temporary, centred around a product launch, and does not aim to build a sustainable mobile development model.
- **Product/brand loyalty**: This involves retaining/rewarding customers by offering them benefits/promises based on the use of the product. Even if the approach tends to qualify the individual based on their interests/tastes, it remains focused on direct ROI, such as the distribution of discount coupons.

Ultimately, none of these isolated approaches aims to build a proximity service-oriented model in a modular manner, communicating anywhere/anytime with open and shared solutions. So, the question was how to streamline all the information as described in Sect. 2.2.

A service-oriented for consumers:
There are many consumer-based social networks, such as travel with 'Tripadvisor.com', products 'amazon.com', restaurants 'thefork.com', etc.

These platforms are a social phenomenon, allowing a new category of informed consumers to buy based on community reviews and ratings. The brands/distributors have understood well that the leads provided by these exchange platforms ensure their profitability. Hence, there are questions on 'neutrality' and product information validation when presenting the product information and whether the real intent is for the consumer rather than benefiting the social networks.

A complex environment:
In France alone, there are different types of mobile devices in circulation with different Operating Systems, and the terminals are reconfigured by mobile network operators with different packaged offers, which distorts the terminal's navigation parameters.

12.3.3 Proxi-Produit Technical Consideration and Implementation—A Project for Extended Packaging Services

If, in a supermarket, one scans the barcode of a product (using the barcode reader available on the shelves), the information presented is only its name and price.

IoT allows brands and producers to present more information to consumers from their barcodes than those printed on the packaging. This additional information is termed as "extended packaging".

This data is present on servers accessible via the Internet. Thus, by using a mobile application that allows the barcode to be scanned and connected to the Internet, consumers can access all the information about the product that interests them. Studies have shown the importance of access to extended product information using a smartphone, both from the point of view of consumers and brands.

Figure 12.1 illustrates the Proxi-Produit service made available to citizens by the French government.

Role of Identifier and Identification Schema in Proxi-Produit:
Identification plays a crucial role in providing citizen-based proximity services. The identifiers that we are used to handling (such as telephone numbers, IP addresses, domain names, etc.) are globally unique. In the same way, the identifier of a product must necessarily be unique.

Thus, the definition and allocation of identifiers must follow specific rules to preserve their uniqueness. Firstly, it is necessary to establish an "identification scheme". Let's take the case of telephone numbers. The allocation of these identifiers follows the "E.164 numbering plan" (ITU-T 1997) as an identification scheme. Each telephone number is limited to 15 digits and includes:

- Country Code (CC)

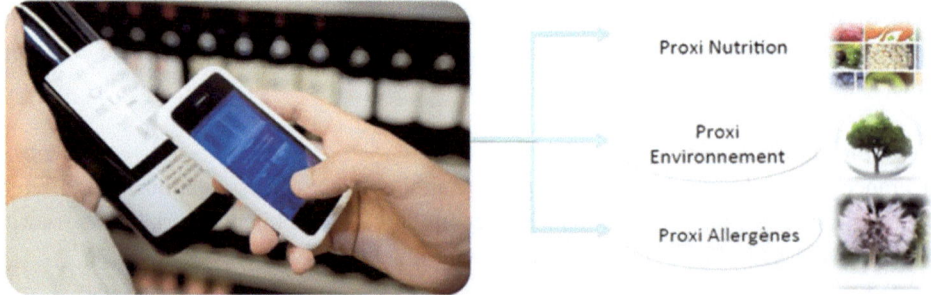

Fig. 12.1 Proxi-Produit—extended packaging use-case for proximity based city services

- National Destination Code (NDC)
- Subscriber Number (SN).

For example, in France, a telephone number following the E.164 numbering plan is 33-1-60760000, where '33' is the CC, '1' is the NDC for the Paris region, and '60760000' is the SN.

Once this identification scheme has been established, it is necessary to define how to allocate the identifiers so that no 'two' identifiers are assigned to the same entity. In the E.164 identification scheme, CCs are assigned according to ITU-T recommendations. NDC and SN are controlled directly by the governments of each nation or the dedicated national organisation. A similar hierarchical allocation scheme is also used in other identification schemes, such as domain names, IP addresses, etc.

The observation we can have here is that in the same way as for the identifiers we use (telephone numbers, domain names, etc.), effective identification of consumer products must follow specific rules:

- It must be based on an identification scheme. Thus, each identifier has the same structure. This allows the implementation and use of a common mechanism for identifier resolution.
- It must be allocated hierarchically. This ensures the uniqueness of the identifiers.
- It must be managed in a distributed manner for scalability and ease of administration.

As the 'Proxi-Produit' project focuses on product identification, there are different product identification schemes. Initially, we started with the Global Trade Identification Number (GTIN) identification scheme used in the consumer goods industry. The GTIN, the unique identifier, according to the GTIN identification scheme, can be used to identify objects globally.

GTINs are allocated hierarchically. Each GTIN is, therefore, composed of two parts (see Fig. 12.2):

- The first, the 'company prefix', allows the company (having manufactured or produced the object) to be uniquely identified ("3112345" in the diagram (Fig. 12.2)). The first three digits of the company prefix indicate which Country the manufacturer belongs to. For example, if the company is based in France, the first three digits are between 300 and 379.
- The second is the 'product code' (67,890 in the diagram above), which makes it possible to differentiate products from the same company (to differentiate two wines produced by the same winemaker, for example).

GTINs are managed in a distributed manner. Company codes are administered at the country level, and locally produced codes at the company level directly.

Fig. 12.2 GTIN barcode example

GS1 as an Organisation assigning and allocating GTIN in the Proxi-Produit project:
GS1 is a not-for-profit, international organisation developing and maintaining its barcode standards and the corresponding issue company prefixes. GS1 has 118 local member organisations and over 2 million user companies. Its main office is in Brussels.

The best-known GS1 standard is the barcode (based on the GTIN identification scheme), a symbol printed on products that can be scanned electronically. The GTIN in the barcode is used to identify a trade item, which could be a product. It's a globally unique number that can help you identify your specific product or service. Figure 12.3 shows the different types of GS1 barcodes and associated GTINs used in the consumer industry.

GS1 positioned itself in the Proxi-Produit consortium as a guarantor in evolving the GS1 standards for publishing information relating to consumer products by brands. It's in the interest of GS1 and its members (manufacturers and distributors) that the services developed consider existing standards (GTINs, barcodes, computerised data exchange protocols, etc.).

GEPIR, the look-up service used by GS1:
The Global Electronic Party Information Register (GEPIR, http://gepir.gs1.org) is a distributed database that provides a unique, internet-based service that allows users to look up information associated with a GTIN. GEPIR proposes to search for information using the GTIN code.

The architecture of GEPIR requires every node to know about every other node. While the number is limited, the maintenance of this list is still very much a manual process. There are other limitations of GEPIR, such as the fact that access to GEPIR is limited for companies that are members of GS1 (Barthel 2014). GEPIR is not an open database or could be interoperable with identifications other than GS1 identifiers. Hence, it cannot

Fig. 12.3 Different types of GS1 barcodes and associated GTIN (*Source* https://www.barcode-us.com/gtins-prefixes-barcodes/types-of-gs1-barcodes)

be used in a service like Proxi-Produit. Figure 12.4 illustrates the resolution of a GTIN identifier to its extended packaging information using GEPIR.

DNS as the look-up service:
Having one global identification scheme (such as GTIN) for all the products worldwide will be nearly impossible. The main reason is that some industries have been using their proprietary coding standards (i.e., identification schemes) for a long time. It is pretty

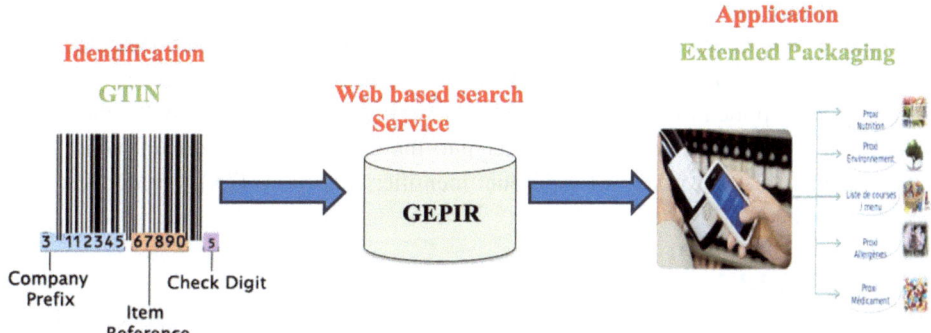

Fig. 12.4 Product resolution using GEPIR

impossible to convince them to move to a newer product identification system. Another difficulty is that it will require consideration of various identification schemes to achieve a global identification schema.

The GTIN allocates the identifiers hierarchically (as explained in Sect. 3.3). The hierarchical nature of such allocations ensures that anyone from any part of the world can uniquely access any service/user in any other part of the world. So, most identification schemes for identifiers in IoT are also hierarchically structured.

The DNS is the look-up service on the Internet, translating "human-friendly" domain names on a TCP/IP network into their corresponding "machine-friendly" IP addresses. Both domain names and IP addresses are hierarchically structured. Hence, the objective was to check whether the product identifiers from the GS1, which are hierarchically structured, could converted to domain names. As mentioned earlier, a similar mechanism exists in ODS, such as using DNS to resolve the object identifiers (their respective identification schemes) to their related digital information.

Proxi-Produit—A fusion of two standards:
AFNIC, as the French Internet registry, manages the Internet domain namespace for France and its overseas territories. Afnic uses DNS, which was developed and evolved as a standard by the IETF (www.ietf.org) for publication and resolving identifiers such as domain names to its information on the Internet. Afnic is positioned in the Proxi-Produit consortium to use the DNS infrastructure to provision and resolve GTINs produced by GS1 Standards via the Internet.

Hence, two organisations working with two different standardisation bodies came together to develop a new standard that could benefit the community and enable openness and interoperability. This new standard is the ONS.

ONS Usage in Proxi-Produit:
In Proxi-Produit, the ONS standard resolves extended information from the product identifier (e.g., barcode). It turns out that the ONS standard uses the DNS (Domain Name System) infrastructure to retrieve information from objects on the Internet. Therefore, the query and response formats must be compatible with DNS, meaning that object identifiers are converted into domain names, and the results correspond to DNS records.

The mobile application scans the product identifier (for example, its barcode with a GTIN identifier) and converts its identifier into the corresponding domain name. Below is the result of this conversion of a product identifier (i.e. the GTIN) to a domain name, according to the ONS standard:

0.9.8.7.6.5.4.3.2.1.1.3.gtin.gs1.id.onsepc.fr
The mobile application retrieves the information related to the corresponding domain name using the DNS resolver. This information is stored in the DNS server as a DNS record. Below is the DNS record corresponding to our bottle of wine:

0.9.8.7.6.5.4.3.2.1.1.3.gtin.gs1.id.onspec.fr. IN NAPTR 0 0 "u" EPC+http "!^*$!http://example.com/!" .

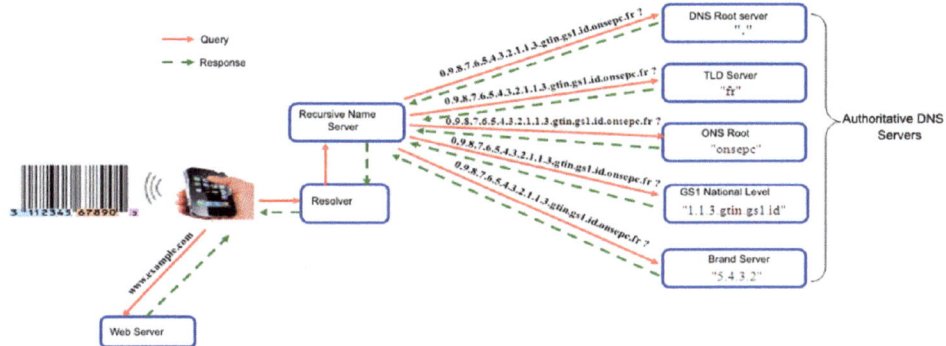

In the absence of cached data, the request is processed by the DNS resolver using the standard DNS resolution process (see Fig. 12.3). The final response from the DNS resolution process will be the DNS record shown above. From the response received from the DNS, the application (on the mobile) extracts the URL (for example, http://example.com/) and requests the appropriate web server (see Fig. 12.3) to obtain the extended packaging of the product. Using the DNS has multiple advantages, which are explained in subsection 3.5.

Security analysis of using DNS for extended packaging in the consumer industry:
As ONS reuses existing DNS procedures and infrastructure (to find information corresponding to the object identifier on the Internet), ONS's security inherits that of DNS. RFC 3833 (Atkins and Austein 2004) contains an excellent summary of DNS vulnerabilities, which (among others) are:

- Modification of IP packets relating to DNS information during transmission and reception
- A malicious person can retrieve the identifier of the DNS request and its type and thus send a false response to the client before it receives the correct one
- DNS "cache poisoning": it is now possible by different means to inject manipulated information into the DNS cache
- Also, a malicious person can compromise the DNS servers which contain the information.

In the 'Proxi-Produit' use case, a patient scans a medical product to find its contraindications. It is possible that a malicious person can hijack the DNS query launched by the application and send back erroneous information, thus leading to a significant risk of jeopardising the patient's health.

Even if the DNS service is, by definition, a highly exposed service, the underlying reason for the majority of vulnerabilities lies in the fact that DNS was not designed to guarantee the integrity of the information that is returned as a response to a DNS query.

The primary approach to addressing these DNS-related security vulnerabilities is called Domain Name Security Extensions (DNSSEC) (Arends et al. 2005). DNSSEC uses public cryptography and DNS data signing. The zone administrator signs the zone file containing DNS data (like the example.com zone).

It is important to note that the signed DNSSEC zone contains the RRSIG and the original unsigned data. The example below shows the contents of record type "A" for an "example.com" zone signed with DNSSEC.

; "A" type Resource Record

example.com. 1 A 192.0.2.1

; RRSIG of the "A" record

example.com. 1 RRSIG A 5 5 1 (20130930064057 20130403064057 3960 example.com.

s8dMOWQjoTKEo1bsK+EYUY+32dsqdsfdsfdsfsdfdssssdfsfsfdsssdfshqt0AaiD=)
When the client makes a DNS query, the signature arrives at the client as part of the response. The client can verify the signature and thus ensure that the response accurately represents the original data. With DNSSEC, authentication and data integrity of DNS responses are ensured.

12.3.4 The Benefits of Involving Two Standards for a Smart City Project Such as Proxi-Produit

The "Proxi-Produit" project has multiple advantages, which are implemented with a cross-pollination of both the GS1 and the IETF standards. Some of the benefits are explained here:

- It enables unique and persistent identification of products that can be shared with many stakeholders consistently and accurately. The product's identity is managed and allocated by a trusted organisation, i.e. the GS1. The information linking the product identifier to its resource is a domain name provisioned by the company, the product owner in the DNS. Hence, the information received by the consumer is trusted information.
- It enables a Top-down approach: for example, in the case of a product recall, the information on recalled products is automatically transmitted to the manufacturer's database. When the consumer "scans" the barcode produced or enters the barcode

number, i.e., the GTIN, the server clearly and simply indicates to the consumer whether their product is affected.
- It enables a bottom-up approach: if a consumer is registered with the application for a risk such as an allergy to certain minerals or a particular product, they will be informed when scanning the product.
- It enables an open standards-based approach: several proprietary product look-up services could be comparable to Proxi-Produit. However, they are not standards-based approaches and are tied to a single service provider. Such alternatives would not fulfil the international business requirements for transparent governance and control over critical network infrastructure.
- It enables dynamic services: 2D bar codes such as QRCode or Datamatrix can encode entire URLs, which are the information associated with the product identifier. This form of encoding URLs, which contains product information in a QR code, is a viable alternative to Proxi-Produit only in cases where the services available for the product are known when the product is commissioned. Usually, the services associated with the product are not known in advance. In addition, many potential web services could interest a user based on context (e.g. Product Authentication, Extended Packaging, Product Recall). Hence, a standardised mechanism must be used to interpret the service type at the URL's end. It is also possible to provide additional services based on contextual information. Some examples are allergen information, language translations, recipes and usage instructions. With Proxi-Produit, it is possible to enable dynamic services based on context, user language preferences, current user location, etc.
- Enables interoperability: Lack of interoperability between different product identifiers has been a burden for the end users. Using DNS as the look-up service for GS1 identifiers enables the possibility of resolving that burden, thus extending its usage to other numbering authorities such as ISO and ITU.

12.4 Conclusions

In a dynamic environment like IoT, where new products and services keep evolving, and network topologies keep changing, automated discovery mechanisms are needed for overall communication management. The discovery mechanism should enable interaction between products or identify suitable services for products not pre-configured or hardcoded as far as the product's address or service endpoints are concerned.

In this article, we explain the Proxi-Produit case study, which satisfies the dynamic discovery requirements of IoT, particularly from a proximity-based citizen service perspective. In addition to technical details, the process involved in designing and implementing the Proxi-Produit platform is explained. We could say that the 'Proxi-Produit' project is the predecessor version of DPP, which was tested in France. The objective of this article is that the experience in the 'Proxi-Produit' project could help us design and develop the tools and technology for DPP.

The objective of DPP is that products sold in the EU will require an attached unique identifier that includes detailed information about materials used, manufacturing processes, recyclability, CO_2 impact, etc. By revealing a product's journey and environmental impact, DPPs will empower consumers to make informed purchasing decisions and pave the way for a greener, more ethical future, thus enabling a circular economy in a Smart city.

References

Amadeo M, Campolo C, Iera A, Molinaro A (2014) Named data networking for IoT: an architectural perspective. https://doi.org/10.1109/eucnc.2014.6882665

Arends R, Austein R, Larson M, Massey D, Rose S (2005) DNS security introduction and requirements. https://doi.org/10.17487/rfc4033

Atkins D, Austein R (2004) Threat analysis of the domain name system (DNS). https://doi.org/10.17487/rfc3833

GS1 Henri Barthel. The need for global standards and solutions to combat counterfeiting, ITU EVENT ON COMBATING COUNTERFEIT AND SUBSTANDARD ICT DEVICES ITU Headquarters, (Geneva, Switzerland, 17–18 Nov 2014)

Gulbrandsen A, Vixie P, Esibov L (2000) A DNS RR for specifying the location of services (DNS SRV). https://doi.org/10.17487/rfc2782

http://gepir.gs1.org

https://www.gs1.fr/

ITFIND—IT 지식포털 (n.d.). http://www.itfind.or.kr/Report01/200611//TTA/TTA-0079/TTA-0079.pdf

ITU-T (1997) The international public telecommunication number plan. https://www.itu.int/rec/T-REC-E.164-201011-I/en?

Langley DJ, Rosca E, Angelopoulos M, Kamminga O, Hooijer C (2023) Orchestrating a smart circular economy: GUIDING principles for digital product passports. J Bus Res 169:114259. https://doi.org/10.1016/j.jbusres.2023.114259

Mealling M, Daniel R (2000) The naming authority pointer (NAPTR) DNS resource record. https://doi.org/10.17487/rfc2915

MISSION PLAN DE RELANCE DE L'ECONOMIE—2009 https://www.economie.gouv.fr/files/finances/presse/dossiers_de_presse/081219relance_economie.pdf

Mockapetris . (1987) RFC 1034 domain names—concepts and facilities

Mockapetris P (1987) RFC 1035 domain names—implementation and specification internet engineering task force

Object Name Service (ONS) (2013) In https://www.gs1.org/standards/epcis/epcis-ons/2-0-1 (2.0.1). Retrieved July 4, 2024, from https://www.gs1.org/standards/epcis/epcis-ons/2-0-1

PROXIMA MOBILE—Le portail de services mobiles pour les citoyens—Bernnard Benhamou (Délégation aux usages de l'Internet—Ministère chargé de l'Économie numérique) January 2013

Proxi-Produit Consortium selected in the Proxima Mobile call for projects. (2009, November 2). Afnic. https://www.afnic.fr/en/observatory-and-resources/news/proxi-produit-consortium-selected-in-the-proxima-mobile-call-for-projects/

REGULATION (EU) 2024/1781 OF THE EUROPEAN PARLIAMENT AND OF THE COUNCIL (2024) In access to the European Union Law (EU-Regulation). Official Journal of the European Union. Retrieved July 4, 2024, from https://eur-lex.europa.eu/legal-content/EN/TXT/PDF/?uri=OJ:L_202401781

Standards | GS1 (n.d.) https://www.gs1.org/standards

www.afnic.fr

www.ietf.org

Sandoche Balakrichenan is the head of Research and Development Partnerships at AFNIC. His research is in networked computer systems, and he is currently interested in IoT identity management, security, and privacy. He has been an invited IoT expert at the European Commission representing the European ccTLD community, IoT expert reviewer at Cap Digital, RIPE IoT Working Group Co-Chair and currently as LoRa Alliance Academic Working Group Chair. He actively participates/contributes to standardisation and associated organisations such as GS1, LoRa-alliance, IETF, RIPE and AIOTI. He received his Ph.D. in Computer Science and Networks from the "Université Pierre-et-Marie-Curie" and is an advisor for PhD students. He is the author or co-author of 21 peer-reviewed articles.

13 Conclusions: Building a Cohesive Future Through Smart City Standards

Okan Geray

Smart city is a relatively novel concept that has been rapidly evolving in the last two decades. Concomitantly, smart city standardization has been at the forefront for cities which embarked upon smart city initiatives.

Standards are commonly agreed upon ways of doing things. They include specific requirements, or they can be general guidelines or best practices. The overall purpose of standards is, inter alia, to ensure consistency, quality, safety, and interoperability. They provide a roadmap for integrating technology in a way that maximizes benefits for cities while minimizing risks.

The development and implementation of smart city standards are critical to creating urban environments that are not only intelligent and interconnected but also sustainable, efficient, and responsive to the needs of their residents. As cities around the world grapple with the challenges of rapid urbanization, environmental concerns, and technological advancements, smart city standards provide a blueprint for harnessing the potential of technology to improve urban living.

The chapters in this book showcase global adoption of smart city standards in different jurisdictions, both at the local and national levels. The following is a list of observations which can be deduced from the case studies covered:

O. Geray (✉)
Senior Digital City Strategy Advisor, Dubai Digital Authority, Dubai, United Arab Emirates
e-mail: Okan.Geray@digitaldubai.ae

- Smart city standards provide a structured framework for cities that guide the deployment of innovative technologies. By establishing clear guidelines and best practices, these standards ensure that technological advancements are integrated in a coherent and effective manner, fostering innovation while maintaining consistency and reliability.
- Standards are essential for ensuring interoperability between various systems used in smart cities. They enable different technologies to work together seamlessly, which is crucial for the integrated functioning of urban services. This integration enhances the overall efficiency and effectiveness of city operations.
- With the increased reliance on digital technologies and data, smart city standards play a pivotal role in safeguarding security and privacy. They provide guidelines for protecting sensitive information, ensuring data integrity, and mitigating cybersecurity risks. This is vital for maintaining the trust of residents and stakeholders in smart city initiatives.
- Smart city standards emphasize sustainable practices and resilience planning, addressing the environmental impacts of urbanization. They promote the use of green technologies, efficient resource management, and resilient infrastructure, helping cities to reduce their carbon footprint and withstand various challenges, including climate change and natural disasters.
- By advocating for inclusive and participatory approaches, smart city standards ensure that the benefits of technological advancements are accessible by all. They encourage stakeholder engagement and community involvement, which are key to addressing the diverse needs of urban residents and fostering social equity.
- Smart city standards adopted include both evaluation and assessment of smart cities (e.g., U4SSC KPIs which eventually became the ITU Y.4903 Recommendation "Key performance indicators for smart sustainable cities", ISO 37120 "Sustainable cities and communities—Indicators for city services and quality of life") as well as implementation aspects.
- The selection of standards for smart city deployments aligns with the specific needs of each urban environment and is tailored to address the distinct goals, strategic objectives, and challenges faced by each.
- Cities' participation in international smart city standards development allows them to articulate their own requirements, expectations, and needs while also exchanging best practices and ideas with other cities.
- Local or national level smart city standardization bodies (e.g., UNE in Spain, NIST in USA) engage with international Standards Development Organizations (SDOs) such as ITU, ISO, IEC to exchange knowledge and to shape smart city standards.
- Some jurisdictions implement platforms to assess and benchmark their cities' performance at the national level (e.g., inteli-gente platform, designed to diagnose, expand, and implement indicators related to smart cities for the assessment of 5570 Brazilian cities).

Smart city standards can unlock several benefits for cities as summarized below:

- The primary benefit of smart city standards is the enhancement of residents' quality of life. Through the deployment of standards based smart city solutions, cities can provide better services and improve the daily experiences of their residents.
- Standards help cities optimize their operations by enabling real-time data collection and analysis. This leads to more informed decision-making, streamlined processes, and reduced operational costs.
- Smart city standards can stimulate economic growth by attracting investment and fostering innovation. The establishment of a clear and supportive regulatory environment encourages businesses to develop and deploy new technologies, creating jobs and driving economic development.
- By promoting sustainable practices and the use of renewable energy sources, smart city standards contribute to environmental conservation. They help cities reduce greenhouse gas emissions, improve air and water quality, and manage natural resources more effectively.
- Smart city standards facilitate the deployment of public safety technologies. These technologies enhance the ability of cities to respond to safety incidents and emergencies, ensuring the safety and well-being of residents.
- Smart city standards accelerate implementation timeframes for cities by capitalizing on commonly agreed standards as opposed to proprietary solutions.
- Smart city standards help in overcoming vendor lock-in and achieving enhanced interoperability.
- Best practices and knowledge sharing through smart city standards help cities save costs, which in turn enhance their operational efficiencies.
- Smart city standards allow cities to compare their performance through commonly agreed and well-defined indicators and measurement methodologies.
- Smart city standards contribute to implementation of United Nations (UN) Sustainable Development Goals (SDGs).

The journey towards smart cities is paved with innovation and collaboration. While the potential benefits are vast, the path forward requires addressing key challenges and staying vigilant in the face of evolving technological landscapes. In this context, smart city standards play a critical role in navigating this exciting yet complex terrain.

Standardization provides a vital framework for ensuring comparability, compatibility and interoperability between diverse cities, technologies, and systems. This fosters a thriving ecosystem where businesses can develop innovative solutions that seamlessly integrate into existing infrastructure. It also allows for the efficient scaling of smart city initiatives, enabling cost-effective replication of successful projects across different urban environments. By creating a common language for data exchange, standardized solutions facilitate the creation of comprehensive data platforms. These platforms empower city officials to make data-driven decisions, optimize resource allocation, and ultimately, enhance citizen well-being.

However, maintaining effective standards in a rapidly evolving technological landscape requires a proactive approach. Regular updates and revisions are essential to ensure standards remain relevant and address emerging technologies in a rapidly changing landscape. Furthermore, fostering international collaboration in standardization efforts is paramount in creating a truly global and connected future for smart cities. By sharing best practices and harmonizing standards across borders, cities worldwide can leverage the full potential of smart technologies to address common challenges like climate change, traffic congestion, and resource scarcity.

In conclusion, smart city standards are not a destination, but rather a continuous journey. They are the foundation for a future where technology empowers cities to function efficiently and sustainably. By prioritizing collaboration, ensuring adaptability, and embracing innovation, cities can unlock the immense potential of standards to create a future where cities are not only smarter, but also more sustainable, equitable, and ultimately, more livable for all residents.

Dr. Geray holds a double major B. Sc. degree in Industrial and Computer Engineering from Bosphorus University in Istanbul Turkey, an M.Sc. degree in Electrical Engineering and a Ph.D. degree in Systems and Control Engineering from the University of Massachusetts in the US. He has published several journal and conference papers and was an adjunct lecturer in management for 15 years.

He has more than 25 years of experience in consulting and advisory roles across various industries. He has consulted for several organizations in Netherlands, France, Italy, Germany, South Africa, Turkey and Dubai. He has worked as the Strategic Planning Advisor in Dubai eGovernment, Dubai Smart Government, Smart Dubai Office and recently Dubai Digital Authority. His responsibilities include Strategic Planning, Strategic Performance Management and Policy Making among others.

Dr. Geray is the U4SSC Chair. He is leading several thematic groups globally for U4SSC namely "Guidelines on Strategies for Circular Cities", "City Science Application Framework", "AI in Cities", "Enabling People-Centered Cities through Digital Transformation", and "Digital Wellbeing". Moreover, he is the Co-Rapporteur of the ITU Study Group 20, Question 7 on "Evaluation and assessment of Smart Sustainable Cities and Communities". He is a member of the IEC-ISO-ITU Joint Smart Cities Task Force (J-SCTF). He has Co-Chaired the Working Group "Economic, regulatory and competition aspects" of the ITU Focus Group on metaverse (FG-MV). He was also the Co-Chair of the "Data Economy Impact, Commercialization and Monetization" Working Group, part of the ITU Focus Group on Data Processing and Management.

MIX
Papier aus verantwortungsvollen Quellen
Paper from responsible sources
FSC® C105338

If you have any concerns about our products,
you can contact us on
ProductSafety@springernature.com

In case Publisher is established outside the EU,
the EU authorized representative is:
**Springer Nature Customer Service Center GmbH
Europaplatz 3, 69115 Heidelberg, Germany**

Printed by Libri Plureos GmbH
in Hamburg, Germany